BETWEEN ALPHA AND OMEGA

What Clergy and Science Professors Don't Want You To Understand

By

Miles Pelton

ISBN: 1-4140-2561-0 (e-book)
ISBN: 1-4140-2562-9 (Paperback)
ISBN: 1-4140-2563-7 (Dust Jacket)

Library of Congress Control Number: 2003098180

This book is printed on acid free paper.

Printed in the United States of America
Bloomington, IN

1stBooks - rev. 10/29/03

Dedication

To the Creator who from the beginning made all things, seen and unseen, with explicit precise built-in provisions for behavior control, including humans who were especially endowed with a brain made in the Creators image with the capacity to learn, reason and to direct behavior in response to that reasoning.

To those who have come to believe that before all else there is a Creator who asks nothing more than to be praised and worshiped before all else whereby, in response to that belief, behavior will be pleasing to the Creator who is God.

To the exposure of Relativism for what it is; an insidious Satan serving belief system that acts through the Creator's built-in behavior control system to confuse individuals into believing that deviant, evil and unjust behavior is acceptable.

To my wife Stella Marie whose steadfast faith in God has been a beacon for my soul.

Table Of Contents

Introduction

For all things there is a beginning. Astounding? Not really. Yet there are many people who refuse to acknowledge such a basic fact especially when our universe is the subject, often resorting to unsubstantiated rationalization and manipulation of word meaning in their effort to deny such truth. Many of those people do not want to know or acknowledge such truth simply because they do not want to deal with what that truth entails. They prefer and are content living in a world where Relativism[1] prevails; where truth is manipulated to fit their preferred intentions. Such is the case as concerns the origin of all things, which is the first truth. Truth concerning any matter cannot be assured except or unless consideration begins with and embodies that first truth, which explains why the Holy Scripture begins with a statement of that fact. It is that narration of the origin of all things that gives meaning and understanding to everything that followed the origin whether revealed in the word or the work of the creator or by divine inspiration.

The motivation for this book began with the realization that truth concerning creation is being deliberately hidden and distorted through modification of the facts so as to accommodate preconceived preferred biased ideological concepts. It became apparent that the distortion of truth concerning creation has caused an escalating ongoing battle for prestige between the community of Scientists and the community of Religious where the sides in this feud are drawn along ideological concepts concerning the role of God, which possibly was the basis for the Cain and Able story in Genesis. In any case, the battle lines have been drawn for or against the concept of creation as narrated in Genesis with both sides in the feud carrying the battle "to the death" while hiding and distorting the truth rather than lose prestige by conceding their concepts are flawed. Why?

In the course of attempting to reconcile the creation argument came a further realization that truth in general, not just on the subject

[1] Where the meaning of words are changed to justify an alternative definition. A form of lying.

of creation, is also being manipulated through a practice called Relativism that has been subtly evolving until it now constitutes a serious threat to the maintenance of a viable society. The practice of Relativism, although existing from the time of Eve in the Garden of Eden, took on a new life when the community of scientists on the one hand united in a conspiracy in their denial of a creator and the religious community on the other hand united in their refusal to embrace knowledge concerning creation and the Creator unveiled by any means other than through their divinely inspired interpretation of the Scripture.

From these realizations developed a belief that sufficient knowledge now exists concerning creation and the processes involved in its subsequent progressive development which, when documented freed of the biases of the scientific and religious communities, would reconcile the truth about creation. With that truth acknowledged, manipulation to serve preconceived biased intentions and the practice of Relativism could be thwarted along with the divisiveness produced by those factors.

As was demonstrated by a recent case involving the word "is", it was made apparent that the intended meaning of words can be altered and laws manipulated through the application of Relativism whereby the intent and purpose of a word is altered or modified to provide a more liberalized or less restrictive interpretation that changes the meaning intended by the originator of a law or document. In that manner, by changing the intended meaning of words used to define a truth as in a law or a document such as the Constitution, the Bible, the Koran or a contract, the intent and meaning of the law, document or contract is modified without changing the wording. There is in fact, a segment of our society that is "making a killing" in legal matters just by applying Relativism to change the meaning of words and that technique has become a frequent tactic of certain political and religious affiliations that pander to a liberal agenda. Throughout history there has been continuing effort to liberalize the definitions of the words used, whether religious or secular, to define behavioral expectations of mankind and it is that practice that is being evoked to hide the truth regarding God's involvement in creation. Consider the impact when the words used by Jesus to define evil behavior[2] are modified to provide a more liberal definition.

One word in particular that has produced great confusion is the word "love." From current definition[3] love means: "**1**. An intense affection and warm feeling for another. **2**. Strong sexual desire for another person. **3**. A strong fondness or enthusiasm," whereas, loved or loving means: "**1.** To feel love for. **2**. To like or desire enthusiastically."

Such definition renders it impossible to establish a meaningful understanding of the term love. What is meant by "an intense affection", "a warm feeling", "a strong fondness or enthusiasm" or even "a sexual desire"? What does it mean "to feel love for" or "to like or desire enthusiastically"? It is quite apparent that such definitions allow almost any desired interpretation and certainly do not define the word as used in Scripture associated with the Judeo and Christian churches. When these definitions are applied in the interpretation of the commandment to "Love the Lord your God with all your heart and all your mind and all your soul" the results do not constitute a meaningful commandment. What does it mean to affirm your love for a spouse in a wedding vow based upon such liberal definition? Is the meaning the same for a possession such as a car or for a neighbor? Do those definitions define how God loves us? By contrast, Saint Thomas Aquinas (1225-1274) in his Summa Theologica, Question 26, Article 4 of the 1st Part of the 2nd Part, opined, "To love is to will the good of another" which establishes a clear, concise, meaningful definition. Would it be unrealistic to abide with such clarity in definition? Does that definition explain what we see as God's love? What is gained by manipulation of the definition if not to accommodate liberal relativistic interpretation of the term love? Since love along with faith and hope are embedded "tools" employed by God to manage the behavior in human beings, is Relativism intent upon subverting the work of the Lord? Is Relativism that proceeds from the insertion of such liberal definitions, evidence of Satan working to subvert the will of God?

---

[2] Mark 7:21-22

[3] The American Heritage Dictionary, © 1983 by Houghton Mifflin Co

A second word of comparable complexity and confusion is the word immoral, which is currently understood to mean: contrary to established moral principles. Unfortunately, there is no clear dictionary definition for what the word means. The Merriam-Webster dictionary[4] defines the word as "**a**: of or relating to principals of right and wrong behavior, and **c:** conforming to a standard of right behavior, or **d:** sanctioned by or operative on one's conscience or ethical judgment."

The listed "definitions" do not establish what is or is not immoral but do acknowledge an impact on behavior. A direct correlation of those definitions with the references to immoral sexual behavior as used throughout both the Old and New Testaments, not to mention the Koran, is not possible, even though immoral sexual behavior is indicated to be a serious behavioral problem. The problem of defining immoral sexual behavior is further confused by the lack of a clear definition for the word sex or sexual. Is there any wonder that immoral sexual behavior has persisted as a major behavior problem since the time of Genesis?

It is not the purpose of this endeavor to convince people to change or take on new religious affiliation. The purpose is to convince people to let known fact speak on its own so that the creation process can be analyzed and behavior-modifying truths established beginning from that point free of biased intentions. In that manner, those processes that are elemental to the behavior of humankind can be revealed whether that knowledge serves religious or secular purposes. The intelligence with which we are endowed says that when it comes to the creation process, the truth as it affects behavior does not differ on the basis of secular and religious beliefs. The truth is the truth for all purposes. In the first place, religion and God exist only with those who believe that the Creator is God but beliefs affect the behavior of people whether those beliefs are based upon that or some other concept. Even Atheists have behavior modifying beliefs. There is even evidence that some pre-human species of life embrace beliefs that influence their behavior. It is a self-evident fact that behavior is

---

[4] © 2001 by Merriam-Webster, Incorporated

influenced by what is believed so a problem arises when behavior-modifying beliefs are based upon other than the truth.

In this work, references are made to Scripture passages as a means of helping in reconciling some arguments because the Scripture is a readily accessible and a commonly used reference for analyzing and teaching human behavior and behavior is a function of human beings with or without the involvement of religion. In fact, the purpose of a religion comes into question if it is not to influence behavior. Also, since truths concerning creation and The Creator God are embodied within and revealed through the study of the work of that creator, those truths should be taken into consideration in the process of interpreting the Scripture. As further fact, the true meaning of the words embodied in the Scripture cannot be understood with certainty if knowledge revealed by the study of the work of The Creator is excluded from consideration or when the meaning of words have been modified through Relativism or interpretation. Points of contention between work and word need to be reconciled and not made into a battle for prestige. It is the battle over the prestige of interpretation of Word versus interpretation of Work that has been fueling the ongoing battle between the Scientific and Religious communities and even between various factions within those communities. It is that ongoing battle that is veiling revelation of the wisdom and knowledge hidden within creation that is fomenting the growth of Relativism as well as radical extremes in religious beliefs. Truth, especially the first truth, needs to be embraced by all and not feared by anyone. If not, any resulting product is suspect.

Working with knowledge winnowed through study of the Word and the Work of The Creator, the source of all knowledge, the author has come to a realization that the truth concerning creation is revealed only when consideration of the matter begins at the logical beginning. Freed of myth, biased perspectives, traditions, manufactured theories and conflicting interpretations, hidden and secreted truths become self-evident which, whether for religious or secular purposes, opens a door to a fuller understanding of the wisdom embodied within the creation effort. Join the author in exposing the wisdom and knowledge held secret in the Word and the Work of The Creator.

Chapter One: The Role Of Behavior

The Omega

For all things there is a beginning and so it is with this universe that is an assembly of minute elemental components and forces. Centuries of research and investigation by renowned Physicists has established that all things in this universe that have a physical presence are an assembly of electrons, protons and neutrons that were themselves made from a single form of energy that has no physical presence. This is a truth. It is this fact that forms the basis of the infamous revelation made by the renowned Physicist Albert Einstein in his $E = Mc^2$ formula. The formula as stated does not, nor did Einstein himself, make allowance for more than one form or kind of energy. The Einstein formula states that all matter, each electron, each proton and each neutron, is made of one form of energy with the amount in each instance in proportion to the amount of matter or mass involved in each of the particles. When this fact is acknowledged and consideration of The Word and The Work of The Creator begins from that point, embedded secreted wisdom begins to be revealed that is freed of ingrained longstanding myth, superstition and theory.

It is the behavioral characteristics of that one common (Universal) energy that has no physical characteristic of which all things are made and the manner of assembly into those elemental particles, called electrons, protons and neutrons, that gives those particles their own unique physical presence as well as their own unique behavioral characteristics. Further, it is forces with no physical presence produced by that energy with no physical presence that not only provide the bonding necessary for assembly of elemental particles but also is the source of all energy in all forms involved in the assembly of the universe. Therefore, *it is the behavioral characteristics of the energy of which all things are made that sets the foundation for the behavior of all creation.* In its basic form that universal energy is seen as the producer of the force called gravitational attraction.

Understanding the creation process from this perspective makes it possible to understand the nature of the Creator and of those things that the Creator has made. It is that understanding that forms the basis for the meaning of all things whether from a secular or a religious perspective. ***It bears repeating. This is an irreducible self-evident truth.*** *All things with a physical presence are made from a single form of energy that has no physical component but possess specific behavioral characteristics, that is assembled into elemental particles that are in turn the building blocks for everything physical in the universe.* The argument of this statement is true whether or not there are any intervening processes involved in converting that single form of energy into those elemental building blocks called electrons, protons and neutrons since electrons, protons and neutrons and the energy involved in their assembly, are the constituent components of all things physical in the universe. That is an established fact. Their method of construction or assembly does not change that fact. That these elemental particles are the building blocks of all things physical in the universe is an ***irreducible self-evident truth***.

All of the particles in each of the three types of elemental particles are identical. An electron now employed in the working of the computer being used to write these words is interchangeable in all aspects with an electron even from the farthest reaches of the universe regardless of the application in which employed. So it is with neutrons and protons. The behavioral characteristics of each type of the elemental particles are eternally identical in each case and there are only three different types of elemental particles used in producing the universe. Further, the assembly of elemental particles always proceeds in building block fashion in a specific order.

The next order in the process of producing physical matter involves the assembly of electrons, protons, neutrons and their related forces, into atoms. The assembly of those elemental particles into atoms always produces a physical component with eternally unique behavioral characteristics that varies according to the number of protons employed in forming the nucleus of the atom. Electrons, protons and neutrons will not assemble except in the form of an atom that possesses a specific eternally unique behavioral characteristic. In fact, a single isolated proton constitutes an atom of hydrogen. More correctly, (as established by Particle Physicists) a neutron that has ejected an electron becomes in fact a proton, which also is in fact an atom of hydrogen. *Each atomic assembly involving a given number of each sub-elemental particle possesses the same eternally identical behavioral characteristics as every other such assembly anywhere in the universe* but a variation in the number or arrangement of sub-elemental particles alters the physical and behavioral characteristics of the assembly. For example, a hydrogen atom consisting of an isolated proton has different characteristics than a proton with an orbiting electron, which in turn has different characteristics than a hydrogen atom with a proton and two orbiting electrons. The important point is that, while each elemental particle used in the assembly of elemental components retains it's own peculiar behavioral characteristics, each unique assembly is endowed with its own unique physical and behavioral characteristics different from the characteristics of the sub-elemental particles and, if disassembled, the unique characteristics of the sub-elemental particles are again discernable. That is, *when electrons with their unique behavioral characteristics are assembled with protons and neutrons with their individual unique behavioral characteristics, the resulting assembly produces a new physical entity with behavioral characteristic that are unique to that specific assembly anywhere in the universe.* Thereby, an assembly of seventy-nine protons, neutrons and electrons always produces an atom of gold that possess behavioral characteristics unique to gold no matter where in the universe that assembly is located. Said in other words, the unique behavior characteristics of the electrons, protons and neutrons are used in this case to produce the behavior characteristics of gold. When an atom of gold is disassembled the elemental particles of which made or the energy those particles are made of, are again independently discernible.

There are several key revelations in the preceding analysis.

First: Energy possessing no physical component produces forces that have no physical component (as in the forces of gravity[5] and electricity) that act upon physical matter to affect the behavior of such physical matter such as to hold electrons, protons and neutrons to form atoms. Those forces with no physical component not only

---

[5] Technically called Gravitational Attraction but for clarity the term Gravity will be used

exhibit their own unique eternal built in behavioral characteristics but also when acting upon physical matter, produce physical reactions.

Second: Each type of elemental physical particle employed in building the universe possesses a built-in behavior characteristic that is unique to and eternally possessed by that type of elemental particle no matter where existing or how employed. *Those elemental physical particles are not only the building blocks with which all the remainder of the physical universe is built, but their unique behavioral characteristics are elemental elements in building the behavioral characteristics of subsequent assemblies as well.*

Third: By combining the physical and behavioral characteristics of the referenced energy and elemental particles, new unique particles called atoms are built which possess new unique built in physical and behavioral characteristics. In each case, *the nature of the physical and behavioral characteristics is determined by the arrangement of the elemental particles used in their assembly.*

Fourth: This process proceeds through each succeeding level of material assembly to build both physical structure as well as behavior characteristics.

Fifth: These revelations that are key to the understanding of all things, including The Creator and creation, are scientific fact recognized by such renowned Physicists as Albert Einstein and by other Particle Physicists. However, rather than incorporating these simple irreducible truths as elementary education, they are being kept hidden and suppressed in favor of theories and myths of questionable authenticity promoted by the scientific, religious and educational communities. **Why?**

Built-in Logic

Embedded within the described process there exists an extremely important hidden built-in capability that is a key factor in the creation of life, that should be discernable to those familiar with basic physics or chemistry as well as those involved in the application of microprocessors. Consider that the electron and proton, the principal ingredients in every particle of matter, are also the principal ingredients in the establishment of electrical potentials. Then recognize that with electrical force, electrons can be moved from atom to atom to produce changes in the behavioral characteristics of atoms (or molecules) including the ability to create a variety of "electrical power situations" such as the "miniature batteries" of ATP (adenosine tri-phosphate) that provide the energy that powers the activity of living cells. Also recognize that electrical force can be employed to produce "physical" forces and thereby movement of physical matter as occurs for example in muscle cells. *Then recognize that atoms of matter, each with their own eternally precise unique **behavioral** characteristics, can be assembled in such a manner as to provide **built-in logic** that can be used to make decisions to control life processes,* for example pulse and respiratory rates. This is a proven truth, not a myth or theory.

Behavior The Objective

By looking back over the beginning or originating process just described, one factor stands out: **BEHAVIOR**. As though the objective in each instance is to establish control over behavior, each elemental element, whether physical or non-physical, and each subsequent assembly of those elements, provides a unique behavioral characteristic. Doesn't it make sense then that behavior is the objective upon which the current "physical world" is dependent, ***another irreducible truth***? Could creation have been successful without eternally precise behavior at each step of the process? The answer is obvious. Therein is the first step, the key to understanding the nature of the Creator God and of nature. The intended behavior of end applications had to have been anticipated and accorded consideration when the behavioral characteristics of those elemental forces and particles were established which had to precede the initiation of their production. It is totally beyond any stretch of comprehension to accept that the behavioral characteristics of those four keystones to creation were established by chance. *That fact alone establishes beyond any reason to doubt that the Creator with unrestricted wisdom has from the beginning willed the behavior of creation.* That is the reason why the elemental particles and energy used in creation possesses the unique behavioral characteristics with which endowed. There is no way that any subsequent action, whether as a "big bang", a "primordial stew" or some other relativistic explanation, could change that truth.

As the creation process was extended through the many levels of increasingly complex sophistication, the ability to control behavior became more and more complex and vulnerable to modification through the application of individual will or malfunction of the embedded processes. With lower forms of animals, behavior was sufficiently controllable with instincts and logic capabilities produced by built in behavioral characteristics. The ultimate test for behavioral control came with the creation of a being whose brain was sufficiently sophisticated as to provide the ability to rationalize, to store knowledge and recall with recombination. In a word, such beings had the ability to learn so as to understand. *To achieve intellect capacity so as to mirror the image of the Creator would require the ability to exercise that capacity without restriction. That is, to exercise freewill.* With that stage of progressive development, reliance upon built in logic, inherited instincts and inherited personality characteristics to control behavior was compromised since such a sophisticated brain could be used at will to overrule built in behavior controls. By creating an individual with the intellect capacity to mirror the Creator's meant that *conscious control over the behavior of such beings had been released to the freewill of the individual* and could not be withdrawn without destroying the species. That is the situation the Lord had to come to grips with following the creation of Adam and Eve even though the initial reaction was that man as created was good. As an aside, is it not a more logical interpretation that the expression "in his own image" as explained in Genesis 1:26-27, referred to intellectual and not physical features since God, acknowledged to exist in spirit form, would have known that he had no physical features to copy? Also is it not the intellectual features, rather than the physical features, that allow man to fulfill the first commandment to be master over all other life forms? Why aren't those conclusions highlighted? Why are they kept closeted in myth and mystery? Would awareness of this irreducible truth not help with understanding?

Evidence of Trial and Error

The narration provided by Genesis Chapters 3, 4 and 6 is very revealing of the chain of events that ensued following the creation of beings whose superior intellect allowed freewill to control behavior rather than inherited instinct and built in logic. Bear in mind that this accounting is claimed to have been made as narrated by God but is seldom if ever covered in religious services. This accounting when taken literally as narrated, clearly gives evidence that the creation of human beings involved considerable trial and error since as narrated, the behavior of humans eventually made God sorry that he ever made them. Using current jargon, God had unleashed a loose cannon.

From the beginning, the **behavior** of these humans was out of control causing the Lord to take some drastic disciplinary actions. Eve fell victim of the relativistic argument of Satan the serpent. The first born of this experiment killed his brother over disagreement on the role of the Lord. Then there was crossbreeding with "other sons of God" that produced giants and other aberrations, and the behavior of mankind was wicked. All of their thoughts were consistently and totally evil. Their behavior made the Lord sorry that he ever made them and the Lord determined to completely wipe out the human race that he had created. But on second thought, the Lord decided to save Noah, whom he found to be a righteous man, the only blameless man alive at the time, his family, and selected animal species. That led to the Ark and the great flood that left only Noah and his family with which to start again the Lord's experiment with freewill. After the waters receded, with Noah and the preserved life forms safe on dry land, God again exhorted Noah to multiply to fill the land, to be a master over all living things whereas God set new rules in an effort to govern behavior by declaring that "murder is forbidden. He declared that animals that kill and people who murder must be killed" as though such direction would compensate for the ill effects possible with freewill.

The narration of events that ensued following the flood also gives evidence that the **behavior** of humans still did not satisfy the Lord. After many more generations, Abram was chosen to be God's example for behavior, because Abram, later renamed Abraham, trusted and had faith in and obeyed God's direction. Abraham was a good example of the type behavior God favored for human beings. For that reason, God entered into a covenant with Abraham in an effort to get Abraham's descendents to behave as Abraham did. As a reward, God promised to make Abraham's descendents a great nation and promised they would be given a large parcel of land but to qualify, each male descendent had to carry a distinguishing mark produced by circumcision. Subsequently, after being rescued by God following 400 years living as slaves in Egypt, God finally levied a series of commandments and instructions upon the descendents of Abraham, then known as Israelites, as a means of influencing their behavior, along with renewal of the promise of reward called a "promised land". In retrospect, giving all of the appearances of another trial and error effort to influence freewill behavior.

The New Plan

Again the experiment failed forcing God, as the Scripture narrates[6], to develop a new strategy to combat the destructive effects that uncontrolled freewill has on behavior. Under this new plan, God, or an extension of God that referred to himself as the Son of God, was given life on earth in the form of a specially conceived child that was called Jesus. From the person of Jesus, God provided explicit and detailed instructions to define acceptable behavior and enhanced validity of his mission by performing miracles including his own resurrection following death. In addition to setting boundaries that would produce pleasing behavior, God, through Jesus offered a new covenant that extended to all people including those who were then not the circumcised direct descendents of Abraham through Isaac and Jacob. The covenant provides that those who *believe* the word presented by Jesus and demonstrate their love of God by obeying the instructions Jesus gave, would gain eternal life after death. ***The key word in this covenant is "believe"***. In his capacity as "The Son of God", Jesus was in fact the savior of the world since the mission of God to create beings in his image would be a failure unless mankind

[6] Jeremiah 31: 31 - 33

learned to control the power of freewill. The instruments that would be used for controlling the behavior of man would be Faith, Hope and Love, the greatest of which is Love as characterized by Apostle Paul. *The act of love is a behavioral characteristic that is produced only when the intellect capacity for freewill decides to "will the good of another".* The greater the intellect capacity the greater is the degree of freewill and therefore, the greater is the capacity for love to control behavior. Some lower forms of life such as dogs for example have some ability to willfully control behavior and thereby demonstrate love for a master but do not have sufficient intellect for freewill to completely control behavior.

When analyzed in this manner it becomes evident that the behavior identified as love, which is the act of willing the good of another, was not a factor prior to the creation of beings with freewill. From there it becomes evident that *love is not an inherited characteristic but is an instrument developed and endowed by the creator to influence freewill choice and thereby behavior.* In other words, our brains were made with an assembly of elemental particles arranged so as to provide a means for freewill, and thereby the behavior of beings made in the Creator's image, to be controlled.

There is one even more significant fact that is revealed by this examination of the glory of God. As explained, prior to the creation of beings in the creator's image, behavior of living things was controlled primarily through instinct and personality traits embedded in and passed along by the genetic DNA structure. Those genetically transmitted provisions provided a rather positive straightforward means for controlling behavior. *Since success of creation was contingent upon controlling behavior, there could be no greater demonstration of faith in and love of humans on the part of the Creator than, after billions of years of work building the universe and advancing the sophistication of life forms, to release control over behavior and thereby the success of those efforts, to the freewill of mankind.* What clearer illustration of God's love could anyone want? From the beginning of time, God has willed the good of mankind. Is that a difficult conclusion? **Why is this kept closeted in myth and mystery?** Is this not a better more powerful illustration of God's love of **all** mankind than is reference to the sacrifice of his only begotten son Jesus since death, even by crucifixion, was not an unusual occurrence? True, that illustration along with the resurrection demonstrated the power of God but does not fully illustrate the depth of his love of all mankind which, as will subsequently be detailed, is the key to influencing freewill choice and thereby the control of behavior. How should I know that God loves me? All together now: by the provisions made in the work of creation so that humans could be endowed with freewill or in other words, freedom.

Chapter Two: The Role Of Freedom

Mixed Messages.

A popular female radio talk show host frequently counsels callers that God is not a puppeteer and we are not puppets on a string. Obviously the intent is to impress the point that we as individuals make the decisions that bring grief into our life and it is up to us as individuals to make decisions that will not bring grief. True, God does not stand over us physically manipulating "our strings" to control our behavior and so we should not blame God when things go awry. This mental exercise is another way of getting the message across that we are ultimately responsible for the results of our decisions. That all sounds nice if we lived in a perfect world but our world and the people in it are not perfect. As was shown in Chapter One, all things in the universe are built of precise elemental particles and energies incrementally in building block fashion. The behavioral characteristics of assemblies are produced when those elemental energies and the elemental building blocks with their unique and specific characteristics are combined into assemblies that acquire their own new characteristics. As the complexity of assemblies grows, so grows the difficulty in controlling behavior. Sometimes things don't always work the way it was intended.

Humans are extremely complex assemblies, especially the mental faculties where the capability exists to support the exercise of freewill. Just the reproductive processes alone involve very complex assemblies and delicate interactions that can be adversely corrupted for a variety of reasons, least of which is incompatible genetic contributions to the process by the parents. From the outset, humans have not always been perfectly produced and some qualities deteriorate over time for a variety of reasons. It is a fact that no individual has control over the characteristics inherited at birth. The combination of characteristics inherited from the parents often produces a wide variation of behavioral characteristics even between siblings. While each individual may be created equal in the sight of the Creator, offspring even of the same parents are not created identically equal except possibly in rare occasions involving identical twins. Each individual is challenged to make the most of what has been inherited, to be all that we can be.

The messages expressed relative to the way creation behaves range from one extreme to the other, neither of which extreme are correct. On one extreme, there are those customarily categorized as liberal who maintain that behavior is a function of "Mother Nature" and individual will. Therefore, our fate is in the hands of an undefined "Mother Nature". There are others customarily categorized as conservative who maintain that God is a "hands on puppeteer" involved directly and "personally" in every aspect of everything that happens in the universe and therefore our fate is a function of divine manipulation. That is to say that no event happens that God did not specifically will to happen. Of course, neither position is or can be totally correct. Both are at least partially correct to some degree in some aspects some of the time. Is there any wonder that people are generally confused? How can we make sense of positions that are partially right and partially wrong at the same time?

The Source of Knowledge

Part of the difficulty in understanding the role of the creator as concerns behavior stems from the fact that knowledge that reveals the creator's role is embedded in the creation process that has been kept hidden as explained by Jesus and recorded in Mark 4:22, "Everything that is now hidden or secret will eventually be brought to light". The role of a creator in the moment-by-moment process of creation is hidden by the complexity of the process and by the employment of energy and forces that have no physical component. (Principally those associated with gravitational attraction) The process of understanding is made even more difficult because creation took place over such a long period of time. Under those circumstances, *an understanding of those things "hidden and secret" is possible only through studying and analyzing the behavior of that which has been created (beginning at the beginning) which is in fact the source of all knowledge.* ***This is a point that deserves repeating***: All knowledge that exists was and continues to be derived through the study of those things that are the product of the creator's efforts, which includes the Scripture as well as the work that went into making the universe. Understanding the process of creation, which is how knowledge is revealed, is possible only when consideration starts where creation began. Refusal of the Scientific Community to acknowledge the involvement of an energy, divine in nature with no physical component, in the creation process on the one hand, in conjunction with refusal of the Religious Community to acknowledge that understanding of creation and the Creator are revealed through study of the creator's work on the other hand, are collaborating to keep the wisdom embedded in those things involving creation that are hidden and secret, from being revealed. The narrow-minded positions taken on this matter by both sides' produces an air of disinterest in getting to the truth and an atmosphere where Relativism thrives.

As detailed in the book by the author, "All Things, Seen and Unseen"[7], and as previously explained, creation had a beginning whether accompanied by a "big bang" or a "primordial soup". Creation began with a specific energy, divine in nature with no physical component that was used to make elemental particles of matter with which neutrons, protons and electrons were made and that are the building blocks of which everything in the universe is made. ***That is the first truth.*** Each of those elemental particles as well as the energy used to assemble those particles, were made to possess individually specific unique eternally precise behavioral as well as physical characteristics from which fact it becomes self-evident that a source of unlimited wisdom, energy, and intelligence is responsible for and is incorporated in those processes. ***That is the second truth.*** When those elemental building blocks are assembled into more complex assemblies called atoms, the new assemblies take on their own individually precise specific unique eternal behavioral characteristic depending upon the arrangement and number of the elemental particles employed in their assembly. *The point that must be understood is that the way these elemental components behave is built in and eternally precise having been set by the behavioral characteristics of the energy from which formed. They always behave*

[7] 1st Books © 2002

the way the creator intended that they behave, from which it must be concluded those characteristics where defined prior to the creation of the first of the elemental components. ***That is the third truth***. Those elemental building blocks set the behavior of all subsequent assemblies built from those components. Their eternally precise behavior sets the behavior of all things including the way our bodies and minds function. While the will of individuals can impact reasoning, it cannot change the behavior of those elemental components. The behavior of the most complex assembly, for example the DNA embedded in each cell of every living thing that provides the instructions governing the behavior of that cell, is set by the behavioral characteristics of the atoms, molecules, and energy used in their assembly and by the arrangement of those elements as assembled. ***That is the fourth truth.*** As the complexity of the assembled particles increases, so too is the possibility that the arrangement of an assembly can be inadvertently or accidentally altered thereby producing behavior different from that intended by design. Such is the case with maladies resulting from genetic mutation.

Signs Of A Conspiracy

As shown by the preceding argument, there has been a progression of development in the creation process that began with the most elemental form of energy and elemental physical building blocks. The process proceeded to produce the basic building blocks now called atoms of matter and then through progressively more complex assemblies to produce more complex compounds of inorganic matter each with a peculiar behavioral characteristic at each step. The process and the behavior of products in this phase of creation are essentially extremely stable. Some would argue to the contrary since certain combinations of elemental particles produce atoms that are unstable and "deteriorate" but which eventually produce a stable state. However, the creation of unstable atoms gives every indication of being a deliberate means of producing otherwise unobtainable specifically desired behavioral characteristics not obtainable by other means.

There are those who are unwilling for some reason to let this type of hidden knowledge be seen. In the words of King Solomon[8], they are like fools who despise wisdom. One extreme insists that creation and the progression of development is an evolution process that happened by chance without the application of wisdom and intelligence. Another extreme insists that the Genesis 1 narration on creation must be accepted literally as written with their own narrow interpretation of the intended definition of words used in that narration, and that the Creator God is in direct command of every event. Since as previously explained, neither argument is either all right or all wrong all of the time, the end effect is to produce a divisive and antagonistic atmosphere that confuses, complicates and interferes with recognition of those things pertaining to creation that are now hidden. Are there ulterior motives? For what reason would those "authorities" that have taken upon themselves the task of ferreting out and revealing truth, take such adamant positions regarding revelation of this knowledge? Would those "authorities" knowingly and purposefully conspire to keep the wisdom that created our universe from being revealed?

Signs Of Built-in Behavioral Control

Analysis of the creation process proceeding from the beginning and progressing in a logical fashion step by step reveals that the behavioral characteristics of the building blocks produces a built in logic that causes assemblies to possess unique behavioral characteristics. The energy of which those building blocks are made also provide the forces for assemblies to react with each other according to their unique characteristics and/or as directed by the DNA in some cases and eventually by the brain. The forces seen for example as producing gravity, electrical energy, light, electro-magnetic radiation, magnetism and atomic energy, exhibit no physical properties and therefore are not directly discernable. Even so, it should not be debatable that a source of energy exists from which

[8] Proverbs 1:7

those forces and energy forms are produced yet the scientific community has refused to acknowledge that fact.

As will be subsequently argued, the energy responsible for producing those forces is directed through the center of gravity of each proton. The force produced causes protons to attract electrons which is the force that holds atoms together, that holds the universe together, that causes muscle cells to contract and that causes electrons to move as in the brain and nerve cells as well as in electrical circuits. When the particles and assemblies of particles are arranged in a particular manner, their built in logic directs and powers their behavior so that a desired action can be provided by an arrangement made specifically to produce that desired action. Instructions for the arrangement of the components employed in the assembly of each living thing is encompassed within the DNA code assembled in the process of fertilization or conception from the maternal and paternal DNA contributions and replicated within each cell of every such living thing. Therefore, the design and thereby the behavior of each individual living thing is preset albeit subject to possible subsequent malfunction due to mutations or as in the case of humans, some aspects of behavior can be "overruled" where allowed by reasoning.

At this point in the analysis of creation, some might and do conclude that all things are made to be freestanding self-supporting without the need for the contribution of a supernatural creator. Of course such a conclusion would be erroneous for several reasons.

<u>First;</u> beginning with the creation of the first particle with physical properties and the capacity to produce force upon another such particle, those particles and each subsequent particle requires a supply of energy to maintain that force as well as their physical and behavioral properties. Since it is totally incomprehensible that those microscopic elemental particles have the ability to store a supply of energy sufficient to last without diminution for the past 12 billion years as currently estimated, it can only be concluded that these particles remain connected to and are supplied energy from their point of origin without which nothing would function. In fact, without that continuing supply of energy those elemental particles of which everything is made would disintegrate. Stated in other terms; nothing

is freestanding. Everything is totally dependent upon the continuation of that supply of divine in nature energy.

Second; all of the features, capabilities and characteristics that exist in the elemental particles are built-in and eternally set according to the design of the creator with their assembly arranged to produce specific behavior capabilities which in living assemblies is controlled by the DNA code written by the creator. So, seen from that perspective, the creator has a "standing" involvement in behavior.

Third; for those humans who believe, the Holy Spirit exists to provide an interconnection with the Lord through a spiritual medium. Some might call that interconnection psychological but the Holy Spirit works to provide an influence on behavior as will be subsequently explained in greater detail.

Who's in Charge?

Preconceived biased concepts interfere with the revelation of those things concerning creation that are hidden and secret which perpetuates the beliefs that either we exist by chance or there are mystical spirits that direct our functioning. Now that more of the hidden secrets have been brought to light, it is easier to understand who is in charge. It is the behavioral characteristic built into the Universal energy of which all things are made and which characteristics establish the physical and behavioral characteristics of physical particles along with the instructions provided in the DNA that makes us what we are. All things were planned, and are built to function according to the creator's specification including the intellectual capability that includes provisions for a conscience as well as reasoning. Our intellectual capacity gives us the ability to reason, to understand, to learn and to remember, to gain the ability to subdue the earth and become master over all life. That behavior characteristic gives us the ability to understand every thing including the ability to recognize good and evil. All of these capabilities are real. They are not mythical religious beliefs or mysterious scientific theory. They have been provided for in our design for a reason and are powered by the energy that proceeds from the center of the universe. An energy that proceeds through the "center of gravity" of each body to and including each proton, which energy supply if interrupted would cause all matter supported to revert to the energy of which composed. That is, matter would cease to exist.

The behavior of the cumulative assembly known as man is ultimately under the direction of the brain (like being the pilot of a plane) that responds to inherited instincts, inherited emotional behavioral characteristics and the physical senses that are built-in as well as to reasoned desires controlled by the individual through freewill. Inherited instincts create within individuals among other things, the desire to be liked and remembered, the desire to live and the desire to learn, the desire to understand. In the process of living, our built in senses cause us to realize that there are experiences that produce pleasing results and there are experiences that produce results that are not pleasing. With the power to reason we come to understand that we can will behavior that produces good or we can will behavior that has the possibility of producing bad results. So, has the Creator God given up control over human behavior? Essentially yes, except when individuals establish (whether knowingly or unknowingly) a faith relationship with the Creator God based upon trust in a belief where upon the conscience is modified to embrace love by willing the good of the Lord our God as well as others, and then behavior reflects that frame of mind which is the Creator's objective. It is important to recognize that the process described is a function of real things with physical properties and serves secular as well as non-secular purposes. This process is not limited to religious intentions. When analyzed in this manner, it is obvious that all things were designed to function with specific behavior characteristics for the purpose of imparting control over behavior, the foundation of creation.

Why Freewill?

It would be presumptuous to infer the ability to know why the Creator made human's with freewill but this examination of the process of creation gives some insight as to the reason.

First: There are the words of Genesis 1:26, "Then God said let us make people in our own image, to be like ourselves. They will be masters over all life—" which infers that the Creator God intended for man to have greater intellect capability than any other living creature plus the ability to direct behavior at will which would be a consequence of the greater intellectual power. Also, the intention for man to become master over life infers freedom to reason and react in response to that reasoning. Precisely why the Creator God wanted to create a being with such intellectual power and freedom may never be known. However, it is heart warming to think that the Creator needs us with our brainpower to help in the management of this world. Insight into the role of DNA and other advancement in medicine give evidence to indicate fruition of that plan is only now coming to light.

Second: There is evidence that the progressive development of life has proceeded from negligible to more sophisticated intellect capability culminating in the human brain. Everything the Creator made beginning with the first elemental energy and physical particle has had complete freedom to behave according to the behavioral and behavioral control characteristics that were built-in. Other than the energy that powers the proton/electron relationship that is the force generator in all things and the inter-acting capability of the Holy Spirit, there are no puppet strings that direct behavior. And so it has been through each step in the progression of sophistication. With that evidence it seems apparent that progression in the development of brainpower, with freedom to reason and act within the capability provided, was at least one of the principal objectives in the development of living beings. To control behavior while allowing freedom so that the intellect capacity of the human brain can be utilized, humans were endowed with a built-in control mechanism that utilizes **beliefs** embedded in a conscience to influence the decision making process.

Third: Reaffirmation of freedom to choose was inherent in the covenant delivered through Jesus. Even though the purpose of the ministries of Moses, who was sent to save the Israelites from slavery, and of Jesus, who came as the savior of mankind, was directed at getting people to love and obey God, the Moses covenants and laws infringed upon man's freedom to reason and act in response to that reasoning. On the other hand, the Jesus covenant does not infringe upon but in fact employees man's freedom to exercise freewill to influence behavior. The Jesus covenant merely says that when through freewill you love the Creator God and believe what he taught through Jesus, your behavior will be pleasing. Again, ***the key word is belief***. It is quite apparent that our creator understood that freedom for individuals to think and to act was necessary if man was to learn, to understand sufficiently to contribute to the extent of the capabilities with which endowed, in promoting the success of creation. It is apparent that even as early as the time of Isaiah the Creator came to realize that the Moses covenant infringed upon that freedom. It is of interest to note that the mechanism for the control of behavior through freewill works even when individuals do not publicly profess belief in God but do believe in a "higher power" or who do respect creation. People behave according to the beliefs that exist in their conscience.

History has shown that when societies permit individuals the freedom to reason with freedom to act on the basis of that reasoning, as was provided by the Creator God, they are more successful and productive. At the same time, history has shown that the more societies restrict freedom to reason with freedom to act on the basis of that reasoning, especially when the restrictions are in response to "anti-Christ" (and thereby anti-freewill) religious beliefs, the less successful and productive that society has been. There can be no doubt that there is a direct correlation between the freedom to reason with freedom to act on the basis of that reasoning and the success of a society. There can be no doubt that the success of a society is adversely impacted where "anti-Christ" religious or philosophical beliefs restrict freedom to reason with freedom to act on the basis of that reasoning. *There also can be no doubt that even though incorporated into the message of Jesus, the "mechanism" that produces the stated phenomena is a characteristic endowed by the Creator to all humans and humans alone so is not a religious thing.* Creation is not a religious thing. Religion is in essence an after thought instigated in an effort to convince mankind to love the Lord God which in turn then influences behavior. The described "mechanism" exists and functions even for those who choose through freewill not to believe in the concept of God the creator or even of a creator.

It should be of interest to note the correlation with the philosophy espoused in the Declaration of Independence, which, based upon the described Creator endowed freewill behavior control characteristic, provides the "mechanism" necessary for the exercise of freedom. Is there any reason to question why "In God We Trust" was chosen as the motto of the United States especially since the inclusion of those inalienable rights in our constitution was not to support a religious cause but was to support the right to freedom including freedom of religion? For the love of God! That's where it came from! The framers of the Declaration of Independence recognized the first truth, creation, and understood the essential elements of the behavior control faculties endowed in all men by the Creator as encompassed within the covenant delivered by Jesus. The United States has prospered because under the Constitution, a "faith contract" is established between each citizen and the governing authority (the Constitution that recognizes a Creator) wherein individual citizens are assured freedom to exercise freewill. When individual citizens trust in the belief that the Creator endowed right to life, liberty and the pursuit of happiness can be achieved through freedom to exercise freewill, their behavior becomes influenced to "will the success of that premise" which in turn produces self motivation to achieve that objective.

Chapter Three: The Heart Of Freedom

Review

In chapter one, it was shown that the elemental particles of matter and the force employed in the creation and assembly of those particles of matter possess unique, precise and eternal behavioral characteristics. It was shown that those elemental particles form the building blocks with which the universe was constructed. It was shown that each unique arrangement of the elemental particles produces assemblies that exhibit there own unique, precise and eternal behavioral characteristics which phenomena is repeated for each subsequent level of assembly. It was shown how freewill was introduced when humans were created and the impact of freewill on behavior.

In chapter two, it is shown that the process known as freewill exists only in species with an intellect capability sufficient to reason and then react in response to that reasoning. It is established therein that humans were the first species endowed with a unique characteristic that provides the "mechanism" to influence the freewill control of behavior.

The Spiritual Heart

The Bible[9] defines the heart as "figuratively, the seat of emotions, thought and intentions, sometimes translated "mind" or "soul"; central or most vital part of something."

The term as defined has been in use since the Scripture was written. In Genesis 6:6 when the Lord became disappointed in the behavior of the people he had created, it is said; "It broke his heart" even though God has no corporeal heart. In 1 Samuel 1:15, Hannah said; "—But I am very sad, and I was pouring out my heart to the Lord." In 1 Kings 11:9 it is said; "The Lord was very angry with Solomon, for his heart had turned away from the Lord—". In

[9] New Living Translation, Tyndale House Publishers, Inc © 1996

Proverbs 24:12 a connection is made between the heart and the soul, where it says; "For God knows all hearts, and he sees you. He keeps watch over your soul, and he knows you knew." The function of the heart is expanded in Isaiah 6:10 where it is said; "Harden the hearts of these people. Close their ears, and shut their eyes. That way they will not see with their eyes, hear with their ears, understand with their hearts, and turn to me for healing." The function of the heart is further expanded in Matthew 15:18 where it is written that Jesus said; "But evil words come from an evil heart and defile the person who says them. For from the heart come evil thoughts, murder, adultery, and all sexual immorality, theft, lying and slander." In Acts 5:3 Peter disclosed that the contents of the heart can be influenced by others when he asked Ananias: "—why has Satan filled your heart?" Obviously the heart under consideration is not the biological heart that pumps blood.

From the preceding references it is clear that there is a function within the human body that corresponds to the functions relegated to the heart. When considered in light of the quoted definition, the function of the heart gives every indication of being associated with the term "conscience". The Catechism of the Catholic Church[10] says; "conscience is a judgment of reason by which the human person recognizes the moral quality of a concrete act." The Encarta® World English Dictionary © 2002 Microsoft Corporation defines conscience as a "sense of right and wrong: the internal sense of what is right and wrong that governs somebody's thoughts and actions, urging him or her to do right rather than wrong." The term "conscience" when considered in this light encompasses the covenant promised by the Lord in Jeremiah 31:33 where it is written; "I will put my laws in their minds, and I will write them on their hearts."

When considered in light of the preceding references, it must be recognized as evident that provisions for the spiritual heart function were incorporated within the design of human beings for the express purpose of providing a control mechanism to serve all reasoned behavior decisions. *The spiritual heart or conscience is a real thing.*

---

[10] © 1994, United States Catholic Conference, Inc.—=Libreria Editrice Vaticana

It exists in and is employed by all human beings in the process of directing any and all reasoned behavior regardless of the purpose of the behavior. It is important to recognize that the spiritual heart or conscience responds to beliefs without regard to the nature of the beliefs or the purpose of the behavior. The conscience works the same for belief 's derived from belief that there is a God for example as it does for belief's derived from belief that there is no God. It works the same for behavior in response to love as it does for behavior in response to hate or evil intentions. The conscience works with complete freewill with but one possible exception, which is the possibility that the Holy Spirit can overrule freewill to influenced behavior. But, if that were a possibility, why is that possibility not used to prevent evil behavior?

The Heart Function

From the preceding discussions it is obvious that the heart function takes place within the mind of each individual and is under the control of freewill. It is further obvious that there exists in memory the ability to establish a "data bank" that includes inherited instincts, personality traits, and BELIEFS formed through experiences, training, education, reason, or imparted through spiritual influences. By functioning in this manner, continuing changes or modifications can be accommodated and stored memory can be overruled by the application of spontaneous freewill reasoning. It is suggested for obvious reasons that the content of the conscience memory bank is in fact the soul or that it constitutes the content portion of the soul.

There can be no doubt that the heart or conscience function takes place within that part of the brain that is controlled by free will. Since only humans have the intellect capacity and provisions for freedom sufficient to control reasoning, only humans have a spiritual heart or conscience. Obviously some portion of the brain in humans is set aside to store the information that makes up the conscience, which is the "heart" of the spiritual heart. Inherited instincts and personality characteristics plus current beliefs and learned reactions stored in the conscience memory aid in the making of behavior directing judgment decisions where the final decision authority remains with reasoning directed at the will of the individual. Initially at birth, the conscience would contain only inherited instincts and personality traits. As things are learned and beliefs established or modified, that information would be stored in the conscience memory.

An important aspect of the spiritual heart system is the ability to modify, add, delete or change the content of the conscience memory through reason directed by the will of the individual. It is this aspect that explains why the covenants with Abraham and Moses were not successful in adequately influencing behavior and it explains why the ministry and covenant by Jesus has been considerably more successful with the potential for continuing greater success. It is a simple concept that behavior is influenced by the BELIEFS held in the conscience memory of the Spiritual Heart, which includes BELIEFS that are established through freewill reasoning. It is what is believed that influences behavior. As Apostle Paul said, "Don't you realize that whatever you choose to believe becomes your master?"[11]

There is a second important aspect to the spiritual heart system that derives from the fact that the superior intellect of the human mind provides the ability for people to establish through reasoning that certain behavior produces good "fruit" while other behavior produces "fruit" that is not good. This system allows that BELIEFS in the conscience that causes the production of "bad fruit" can be changed so that "good fruit" can be produced. This is a message often repeated by Jesus. It is also the basis for his assertion that salvation would result when people believe in him and obey his word because by so doing beliefs that produce "good fruit" would be established in the spiritual heart or conscience. The same reaction is involved with the commandment to love the Lord your God since to do so would result in the establishment of a BELIEF in the conscience that would produce behavior acceptable to the Lord. In both cases, beliefs that produce evil behavior would be suppressed in or crowded out of the spiritual heart.

---

[11] Romans 6:16

Examples of The Heart at Work

Consider the following scripture passages in light of the previous explanation of the spiritual heart system for an enlightening basis for the passage:

- Deuteronomy 5:6; "Love the Lord your God—." When this mandate is believed it will install beliefs in the heart that will produce a positive behavior influence.
- Deuteronomy 5:7; "Do not worship any other gods but me and do not make idols of any kind." The worship of other gods or idols would insert beliefs in the conscience that would take the place of beliefs that would produce good influences by loving the Lord.
- Matthew 22:37; "—you must love the Lord your God and love your neighbor as yourself." When the conscience is occupied by belief in the Lord the reasoning process would be influenced to choose good behavior and would crowd out those beliefs that influence bad or evil behavior.
- Matthew 15:19, "—from the heart comes evil thoughts" which is essentially a repeat of Proverbs 23:7 (King James Version) where it is said "For as he thinketh in his heart, so is he"[12]. It is the beliefs recorded in the conscience of the spiritual heart that influences the thought process of individuals.

Spotlight On Satan

Throughout the history of Judeo/Christian religion, the status of Satan has been glorified and ranked in power comparable to that of the Lord even though wrapped in mystery. According to the Gospels,

---

[12] It is of interest to note that the more recent New Living Translation states, "Eat and drink, they say, but they don't mean it", which conveys a vastly different meaning, an example of relativism at work.

behavior categorized as evil was attributed to the existence of demons and evil spirits ***that lived in people***, which were in turn understood to be under the domain of Satan. The Gospels document that the "casting out" and the curing of people infested with demons and evil spirits played a significant role in the ministry of Jesus and in the ministry of the early Apostles. There can be little doubt that the role of Satan, the devil, demons and evil spirits was a major concern of Jesus. The description of those documented cases indicates some considerable misunderstanding on the matter since some behavior that is now known to be produced by medical and psychological problems, where then attributed to demons and evil spirits. Even so, it is of particular interest to note that since the advent of Jesus, incidents with acts attributed to demons or evil spirits have diminished. There is good cause for that sudden change. It has been demonstrated repeatedly that when beliefs recorded in the conscience that produce evil thoughts and behavior are "cast out" and replaced with beliefs that produce acceptably good behavior, for example, as are produced through belief in Jesus, the incidence of evil behavior diminishes. That diminution in incidence goes in line with the growth of Christianity.

As previously related, from the time of Abraham to the time of Jesus, faith contracts with God were established with godheads such as Abraham, Moses, King David or the prophets. Faith contracts were not established between individuals and God. Instead, individuals were charged to obey Commandments wherein the individuals were exhorted by the principal Commandments not to worship any other gods except the Lord and not to make idols. As a reward, those who complied without fail would gain access to the Promised Land. Since the command was to worship, faith contracts based upon loving God were not established between individuals and God so that beliefs based upon the love of God were not established within the conscience of their spiritual heart. In Romans 9: 1 Apostle Paul made that point where he said: "Now in that first covenant ***between God and Israel***, there were regulations for worship and a scared tent here on earth." The absence of a belief by each individual Israelite (or anyone else for that matter) in the need to love God left a void in the conscience of their spiritual hearts. That gave free reign to behavior guided only by freewill reasoning influenced by idol worship, evil spirits, the love of personal pleasure and the practice of Relativism. It was the filling of this void in the conscience with beliefs that did not condemn evil producing idol worship and the attainment of personal pleasure at any cost that gave rise to the concept of a Satan. Surely, since our spiritual heart uses the content of the conscience to influence freewill control of behavior, whether established by belief in the need to love God or any contrary beliefs seen collectively as Satan, it would be seen that ***Satan has power equal to the power of God*** when it comes to influencing the behavior of humankind. It is important to recognize that it is the belief contents of the conscience that influences behavior. Both God and Satan work to control the beliefs embedded in the conscience and thereby to influence the behavior of individuals. By definition then, the thing called Satan would be the cause for evil producing beliefs being stored in the conscience of an individual. Realizing that Satan functions through the conscience of the spiritual heart and is only ***the name given to evil producing beliefs***, Satan did not exist or was not a factor prior to the creation of human beings since humans were the first to be endowed with the intellect and reasoning capacity needed to direct behavior through a spiritual heart. By the same token, the phenomena called Satan will be around as long as there are human beings.

Relativism and Satan

Relativism is the phenomenon identified by the phrase "Defining Deviancy Down" coined and made popular by the late Senator Daniel Patrick Moynihan (D, N.Y.). It is the practice of applying definition to a word or situation based upon a perceived relevance that produces a response less demanding on the perceiver than would result otherwise. It encompasses behavior sometime referred to as Liberalism. The application of Relativism acts to change the definition of terms embodied within and critical to defining the conditions of a contract or agreement specifications and thereby changing the intended meaning of the contract or specifications. When applied to the content of the beliefs held in the conscience of the spiritual heart, the intent of terms such as those defining greed, deceit, sexual immorality, theft, slander, and murder for example (whether in response to belief in Jesus or because of civil law), are modified so that such behavior is seen in their own minds as acceptable and thereby does not instill fear of condemnation.

When Relativism is seen in this context where its application modifies beliefs held in the conscience that normally would warn against evil behavior in the same manner and to the same degree as any action of Satan, it is apparent that ***Relativism is not serving the cause of God but the cause of Satan***. Does that not fit the definition of sin? Since Eve succumbed to the "relativistic" argument of Satan, does that make Relativism the "original sin"? You decide. Also, *since it is shown through examination of the creation process that the behavior of the materials of construction at each level is extremely precise with no room for relativism, would that explain why those who practice Relativism have set their minds against acknowledging creation and evidence of an intelligent Creator?* Is that further evidence that Relativism is the "original sin" that causes people not to see or hear or understand? Further, *since Relativism functions in the mind to remove the stigma from "evil behavior" similar to the reaction to "mind altering drugs," is that why so many are so willing to embrace Relativism?* It sure looks that way since under the influence of Relativism people (including pastors) sometimes do things that they would otherwise not risk.

The Origin of Relativism

The "mechanism" that permits the practice of Relativism has its origin in the mind of human beings. It is the greater reasoning power of the intellect of humans made in the Creator's image that gives rise to freewill and that opens the door to the practice of Relativism. That mechanism is produced by the physical arraignment of the mental faculties existing in humans. It is latitude available to all human beings which gives rise to the mystery of why some but not all people succumb to Relativism? Does this explain the mystery of why some people have eyes to see but do not see, ears to hear but do not hear and hearts that cannot understand? Is there a quirk or defect responsible? Consider that the inherited personality traits of each individual also are embedded in the conscience to influence freewill behavior decisions. Could it be those inherited characteristics that determine who will be or who will not be susceptible to the practice of Relativism? That certainly would be a factor at least and comes close to mirroring the reality of the decision-making processes involved. There is no question but that through inheritance, some people are inclined to be more sympathetic than others and there are indications that those with an abundance of sympathy also have a tendency to favor or even misdirect their sympathy toward liberal agendas through the practice of Relativism. Bear in mind that sympathy is not the only inherited personality that impacts on freewill behavior decisions. Some people inherit a tendency to be mean or rebellious or selfish for example. The point is that the beliefs embedded in our conscience must be adjusted to give preference to loving God before all else and our fellow man as ourselves.

These are difficult observations to verify since inherited tendencies can be influenced in either direction through education and will. Does that mean that an effort is needed to control behavior by controlling heredity? No. It means we need to do a better job of identifying and teaching the truth beginning with the first truth that all that exists is the product of an intelligent creator. But, that leads full circle to the mystery of all mysteries, how to get people to see, to understand?

Chapter Four: The Love Factor

How Can One Know Love?

Through the ages, children have been taught a ditty that goes something like "Jesus loves me, this I know. For the Bible tells me so". Of course there is no mention of their name in the Bible even though it is said that God knows all by name. There is no record in the Scripture of either God or Jesus making a specific claim that they love us. However, even to this day, ministers, clergy and counselors are frequently heard assuring people with the statement that "God loves you." Such assertions are often of questionable value and in some cases, best not given but might have greater impact if a reference could be given to show where God or Jesus spoke the words "I love you" or how one is to know that God loves us. In a word, show some proof above and beyond a third party claim.

Throughout the scripture various people have given similar assurances or have expressed belief that God loves or has loved them. The closest words to that effect appears in John 13:34 where Jesus speaking to the apostles, says: "So now I give you a new commandment: Love each other. Just as I have loved you, you should love each other", but even those words fail to define what love is. So how is one to recognize God's love or know that God, or anyone else for that matter, loves us? How are we to know what is involved in loving someone?

The dictionary/concordance of the New Living Translation Holy Bible provides this convoluted definition of love: "the ultimate expression of God's loyalty, purity, and mercy extended towards his people—to be reflected in human relationships of brotherly concern, marital fidelity, and adoration of God." Give me a break for the love of God. Surely we can provide a better definition for love than that.

In 1 Corinthians, Chapter 13, Apostle Paul went to considerable length in an effort to define love. He described some of its characteristics and declared that love is an important virtue that lasts forever. In his words, "there are three things that will endure—faith, hope and love—and the greatest of these is love." But he did not define love and the Scripture otherwise is essential silent in that regard.

The Catholic Encyclopedia at *www.newadvent.org* references the 1 Corinthians passage and states that love is usually called charity defined as "a divinely infused habit, inclining the human will to cherish God for his own sake above all things, and man for the sake of God." The classification of love as a habit is to say the least, very confusing.

The American Heritage Dictionary of the English Language, Fourth Edition, 2000, defines love as: "**1**. A deep, tender, ineffable feeling of affection and solicitude toward a person, such as that arising from kinship, recognition of attractive qualities, or a sense of underlying oneness. 2. A feeling of intense desire and attraction toward a person with whom one is disposed to make a pair, the emotion of sex and romance. **3a.** Sexual passion. **3b**. Sexual intercourse, a love affair. 4. An intense emotional attachment, as for a pet or treasured object. 5. A person who is the object of deep or intense affection or attraction, beloved. Often used as a term of endearment. 6. An expression of ones affection. **7a**. A strong predilection or enthusiasm. b. The object of such an enthusiasm. 8. Love (Mythology) Eros or Cupid. 9. Often Love (Christianity) Charity. **10**. (Sports) a zero score in tennis." These are the currently popular liberal definitions that satisfy the purposes of Relativism since there is no clarity of definition. They could hardly provide an explanation of what Jesus meant when he said Love the Lord Your God.

The Catechism of the Catholic Church[13] defines love as a passion although it also references the work on the subject by St. Thomas Aquinas who concluded: "To love is to will the good of another". This is an especially insightful conclusion since by that definition love must derive from the will of the lover, which demands a behavior that reflects a frame of mind or a belief in the conscience. A verbal declaration of love alone does not satisfy that definition. The reaction to the act of willing the good of another remains with the lover wherein the resulting frame of mind causes behavior that shows that

[13] Liguori Publications, English translation copyright © 1994

frame of mind to the one loved. Love is therefore not an inherited instinct or a declaration in pursuit of person gratification. The object of one's love can be known solely through the behavior that is produced because of the frame of mind produced by a belief in the conscience of the lover.

While the Merriam-Webster Online Dictionary includes the currently popular definitions, it also includes the Thomas Aquinas concept as the fourth option. Merriam-Webster says: "unselfish loyal and benevolent concern for the good of another as **(1)**: the fatherly concern of God for humankind **(2a)**: brotherly concern for others or **(2b)**: a persons adoration of God." It is not known how long Merriam-Webster has been carrying this option but it is heartwarming to see such a meaningful definition in this respected dictionary.

A Definitive Definition.

The Aquinas definition now included by Merriam-Webster is especially gratifying since there is now at least one dictionary that provides at least a choice of a relative definition that defines how God loves as well as how individual humans love. By the Aquinas definition, we will know that God loves us because of his behavior over the billions of years of effort to produce an environment and life support essentials necessary for our existence as well as the features that make us in his image. Over all of those years, God willed our good, which is to say, God showed his love. It would be absolutely illogical to conclude that any of the other definition choices explains God's love for us or clarifies the way God intended that we should love him. How should we love God? According to the Aquinas explanation and as commanded by Jesus, we should will good of God's objective, creation, with all our heart, with all our of mind and with all of our soul, which, if we do, will be shown by our behavior. The Aquinas definition also provides clarification of the commandment to love your neighbor as yourself since it means that we should will good of our neighbors in which case our behavior would reflect that frame of mind. When considered "in to-to", there is no need for any other than the Aquinas definition.

The Role of Love

Certainly the currently accepted dictionary definitions for love do not make sense when applied to the term as used in the Scripture when Jesus said "You must love the Lord your God with all your heart, all your soul, and all your mind", except the definition offered by Aquinas. The confusion resulting from this comparison of definitions result from the lack of understanding concerning the purpose of love and how love functions.

When the definition offered by Aquinas "to will the good of another" is acknowledged, the meaning of love begins to become clear. When recognized that love influences the behavior of the lover by working through the heart or conscious of the lover in the process of applying freewill to decision making, it becomes self-evident that love is a tool endowed in humans at the time of creation to help influence behavior. The purpose shows more clearly when love is applied as in a faith contract involving the trust in a belief, such as in the belief in God. When an individual believes in God and trusts in that belief, that person will "will the good of", which is to will the success of, God's objectives. That love will be impressed upon the heart of the lover where upon it will work through the conscience to influence freewill decisions of that individual toward the intention to will good. The greater the trust in a belief (FAITH), the stronger the resulting love will be which will increase the influence on behavior.

Love does not work only with the belief in God. The same reaction is produced when willing the good of a friend, a neighbor, a spouse or even a possession such as a home or car. In each case, through willing the good of that person or belief, the attitude of the lover is influenced and that influence is reflected in the behavior of the lover. The process becomes convoluted however when a person gives precedence to willing the good of self, or of things that give self gratification, greater than the precedence given to willing the good of God above all else and ones neighbor as ones self.

Following this analysis it should be evident that matters such as sex, and the other behaviors defined by current dictionary definition are not love in the sense of willing the good of another. Gratification from sex derives from the instinct to reproduce. An individual may for example engage in sex when the belief that is influencing that behavior may be the expectation of personal gratification separate and apart from the love, that is willing the good of, another. However, there is no more powerful behavior motivator than when love and the instinctive desire for sex are conjoined.

Discussion of this subject cannot be considered complete without a word of warning, especially for adolescents but not for adolescents alone. Love, willing the good of another including the love of God, is easily and frequently faked. We become so anxious to "feel loved" that our impatience often makes us easy victim of pretenders who use the pretence of love to take advantage of those who are not wary. That same urge to experience the feeling of love is often responsible for reasoned modification of beliefs embedded in our conscience. Such was expressed by the words of Proverbs 23:7 (KJV) where it was said: "As he thinketh in his heart, so is he."

Chapter Five: The Issue of Faith

The Faith Phenomenon

While it is hard to find a consensus of opinion on just what faith is, there is agreement that faith is a phenomenon that is critical to the functioning of the society of human beings whether in secular or non-secular activities. In the secular arena, faith is a key factor in the ability of our modern day society to function. Our society today could not function if people did not believe and then have faith in the reliability of equipment they depend upon for commerce and every day activities such as mechanical refrigeration and traveling by car or airplane. Apostle Paul picked up on the concept that faith is the key factor in the formation of a special relationship between man and God in Romans 4:13—16. It is interesting to note that prior to the ministry of Jesus, except when dealing with the relationship between God and Abraham, the terms belief and faith were not emphasized in explaining the relationships that should exist between men and God. It is however, recorded in Isaiah 43:10 where the Lord laments through Isaiah: "You have been chosen to know me, believe in me, and understand that I alone am God". This is not to be inferred that at least some of the people of the Old Testament days did not have faith in God although it is obvious that God was disappointed with the behavior of Israelites. It is in the ministry of Jesus that the term "**belief**" was revealed to be the key factor in the development of faith in the promise of eternal life wherein the believer trusts in the word offered by Jesus. It is that trust in that belief that is the foundation upon which Christianity is built. Faith then is the instrument by which the creator God influences behavior in those beings equipped with freewill. As will be seen, faith is the act of trusting in a belief whether in secular or non-secular activities. It is through faith born of trusting in a belief, that we "will the good of" or love another.

Multiple Facets of Faith

Faith does not exist alone. As will be shown, the formation of faith requires the presence of a number of different elements foremost of which are belief in, trust of, and obedience to a set of tenants. Upon examination it is seen that the "facets" involved constitute a contract. A typical example of such a set of tenants is the Christian Profession of Faith, however; the elements and the tenants involved in a faith contract are not always so explicitly defined.

In Genesis 15:4-6, the Lord declared Abraham righteous because of his faith after Abraham believed the Lord's promise that he and his descendants would prosper and "would become too many to count". Later, Chapter 22 recounts how God tested Abraham's faith and obedience by demanding the sacrifice of the son long sought and necessary if there were to be descendents "too many to count", the fulfillment of which promise required a male heir. The Old Testament is replete with accounts of "faith based contracts" even though not specifically identified as such. Surely Moses reacted in response to faith in the promise of the Lord to assist him in leading the descendants of Abraham from slavery in Egypt to the Promised Land. The contract required that Moses believe and then trust in the promises of the Lord which lead Moses to comply with the tenants of the Lord's plan. It is of interest to note that the plan to free "his people" from 400 years as slaves in Egypt was not achieved by direct action by the Lord but through the invocation of a "faith contract" with Moses. Even though of apparent extreme importance, the Lord put his trust in that "faith contract" which leads to the conclusion that whether by design or necessity, using people to accomplish his objectives is the method of operation employed by the Lord. Some of the facets involved in faith are seen clearly in the preceding examples. Certainly Moses of his own freewill, since the Lord did not use force or coercion, believed and trusted the Lord and by his behavior demonstrated, according to the Aquinas definition, his love of the Lord. So faith in the case of Moses involved the freewill application of belief, trust and love.

There is a concealed aspect in the Abraham and the Moses faith contracts that was not immediately but subsequently became apparent to the Lord. In both cases, the Lord endeavored to use faith contracts with Abraham and subsequently Moses as a means of evoking the "righteous behavior" of their subjects without establishing faith contracts independently with each individual Israelite. Since those faith contracts where between Abraham and the Lord and between Moses and the Lord, the remaining individual Israelites were not "bound" in their own faith contract with the Lord and so their behavior was not materially influenced. That plan failed since the Abraham and Moses covenants did not establish a faith contract between each individual Israelite and the Lord and that is the key to eliciting freewill obedient behavior of individuals. Further, the imposition of restricted behavior under the Abraham and Moses plans as documented in Leviticus and Deuteronomy, restricted and limited freewill. In other words, when trust in a belief that there is a God the father and creator of all things, seen and unseen has not been established, the individual is not moved through faith to behave in a manner that reflects love of God but may be in fact vulnerable to being influenced by other possibly contrary beliefs.

Faith and the Ten Commandments

The advent of Jesus marked a major new approach in the manner with which God planned to elicit the righteous behavior of mankind. It is important to recognize that rather than restrict or limit freewill so as to control the behavior of mankind, as was the case with the Moses covenants, God's new plan was conceived to elicit righteous behavior voluntarily by establishing faith contracts independently between himself and each individual. Apostle Paul recognized this difference when he said: "That is why he mediates the new covenant between God and people, - - -."[14] Prior to that time God had employed the promise of a reward along with the fear of his wrath on the Israelite populous as a whole, in his effort to solicit their righteous behavior. That method produced a faith relationship between the principals Abraham, Moses, and David and the individual Israelites rather than between God and the individual Israelites, where upon those leaders

[14] Hebrews 9: 15

became idolized instead of God. The Old Testament documents the failure of those efforts. Apostle Paul recognized this defect in the Moses faith contract when he said; "If the first covenant had been faultless, there would have been no need for a second contract to replace it."[15] "When God speaks of a new covenant, it means he made the first one obsolete. It is now out of date and ready to be put aside."[16] Then Paul followed with the observation that the first covenant was between God and Israel[17] and not between each individual Israelite and God.

The objective of God's new plan, initially directed at the descendents of Abraham but subsequently expanded to include all people, was to get people to recognize, acknowledge and believe that Jesus, sent as a human, carried the word and authority of the creator, the God of their forefathers. In that role, Jesus with a physical form, made God visible to the people. During his ministry on earth, Jesus gave the world "The Word of God" and explained the behavior that would be pleasing to God including a reaffirmation of the commandment to love God, not himself, above all else and "to love our neighbors as ourselves". Through him God performed miracles, including resurrection after death, to emphasize his relationship with God. In addition to leaving us "the word of God" Jesus offered a new covenant that those who believe that he was who he said he was and obey "the word of God" that he delivered, will gain eternal life.

Jesus obviously understood there would be a problem getting the Israelites to understand his role as an extension of God. He repeatedly, explicitly and emphatically redirected to God, the Father of all, all attempts to glorify him. Even so, the tendency to idolize Jesus apparently developed and persisted to the extent that resolution became the principal subject considered in the first (325 AD) and second (385 AD) Ecumenical Councils. Use of the terms "son of man" and "son of God" did not help much in clarifying the role of Jesus. Surely by now we are sufficiently educated as to be able to understand and accept the pleas of Jesus in this regard. According to

---

[15] Hebrews 8: 7
[16] Hebrews 8: 13
[17] Hebrews 9: 1

Jesus, it is with God the creator that we need to establish a faith contract.

The Role of Faith On Behavior

As explained in Chapter One, the process of creation involves the application in building block fashion, of energy and forces that have no physical component yet are able to manipulate objects that do have a physical composition as well as to manipulate other energies and forces that themselves do not have a physical component. Some of these actions produce what are called psychological reactions. On this matter, it must be understood that as the creator of all things, seen and unseen, the creator God is the master of psychological forces as well as other unseen forces such as gravity that are more readily discernable.

When an individual believes in Jesus (as God not as a man) and his message and trusts in that belief, a unique faith contract as previously explained is thereby established between the individual and God. Under those circumstances the frame of mind of the individual is inclined in response to that faith to will the good of God and God's interests and then that frame of mind influences the behavior of the individual. The magnitude of that faith is dependent upon the degree of trust in the content of the belief that forms the basis for that faith. When the individual develops faith by believing and trusting in God, the resulting frame of mind will influence benevolence toward God's interests which is to influence people to behave within prescribed limits without restricting freewill. As with Abraham whose faith was tested by God, obedience of God's will as a result of willing the good of God is the test of faith.

It is the faith developed within the described faith contract that Jesus spoke of so frequently while expressing "the word of God". He admonished people when their behavior reflected little faith because their trust was not complete. He performed miracles for people whose behavior demonstrated great faith. In Matthew 14:31, Jesus admonished Peter when he failed to walk on water after Jesus told him he could by saying: "You don't have much faith. Why did you doubt me?" In Mathew 7:10-13 Jesus admired the faith of a Roman officer who asked that his servant be healed. He told the Roman officer: "Go home. What you have believed has happened." Trust in a belief is embedded in each of these references.

It was Apostle Paul who picked up on the relationship between faith and salvation. His "theory" that is debated, misinterpreted and responsible for producing a rift in the Christian Church, was developed in Romans 3 where he, schooled in Judaism was addressing people who were likewise familiar with Judaism. He recognized that no one could be made right with God by doing what the laws given through Moses commands because those laws were not being obeyed for reasons previously explained. Paul had experience to back up that argument. He realized that with the advent of Jesus, "we are made right in God's sight when we trust in Jesus Christ to take away our sins."[18] (That is when we trust in a belief.) And he said "We are made right with God when we believe that Jesus shed his blood, sacrificing his life for us."[19] It was Paul's conclusion that since God did not punish those who sinned under the old law while declaring those who believed in Jesus to be right in his sight even though guilty of sin, that "our acquittal is not based upon our good deeds. It is based upon our faith. So we are made right with God through faith and not by obeying the law."[20] (Again, faith is trusting in a belief.) He was speaking of the law passed down through Moses. He continued in 3: 30: "—there is only one way of being accepted by him. He makes people right with himself only through faith—." (When we believe Jesus and trust in that belief.) But then Paul concludes in 3:31, "Well then, if we emphasize faith, does that mean that we can forget about the law? Of course not! In fact, only when we have faith do we truly fulfill the law." It is important to remember that by believing Jesus and then trusting (a freewill act) in that belief, the resulting faith will produce obedient behavior. That is the point made in the Letter of James where James said: "Faith that doesn't show itself by good deeds is not faith at all. It is dead and useless."[21]

---

[18] Romans 3: 22
[19] Romans 3: 25
[20] Romans 3: 27-28
[21] James 2:17

James followed with: "I can't see your faith if you don't have good deeds, but I will show you my faith through my good deeds."[22]

The position of Paul on the relevance of faith has been and even is today, sadly misunderstood and misinterpreted. Some Relativists see the statement that "So we are made right with God through faith and not by obeying the law" to mean that all we need to obtain salvation is to profess faith in Jesus. That is the relativistic type of argument that Satan would make. Following that logic would mean that our behavior is of no concern to the creator and our fellow man, which is not what Paul said. It is clear that Paul understood that believing and trusting in Jesus are prerequisites in the development of faith and that when faith is produced by satisfying those prerequisites, we will of our own freewill endeavor to live in such a way as to produce good deeds while fulfilling the law which under the new Jesus commandment consists of loving the Lord before all else and our neighbors as ourselves.

Faith Transcends Religion

The process where faith influences behavior is not limited to faith contracts with God but applies to all situations, including marriage, where individuals possessing freewill capability trust in a belief. Also, to understand the concept, it must be recognized that the trust pertains to ***the content of the belief*** that sets the conditions of a faith contract. Faith cannot be established by words as in a profession. Faith is the product of an action based upon trusting in a belief. Consider the case of a parachutist. The parachutist first proclaims that he believes that the parachute will work. That sets the content of the belief. Then the parachutist trusts in that belief when he agrees to jump from a plane but it is through the act of jumping with the parachute that faith is demonstrated. Faith would not have been involved if the parachutist had not demonstrated a trust in that belief by jumping.

Similar faith contracts exist when we trust the reliability of automobiles or even refrigerators. Without such a faith contract, we

---

[22] James 2:18

would not likely consent to ride in an automobile traveling down a highway or depend upon a refrigerator to protect our perishables. We consent to do those things because we believe and trust just as you would believe and trust what Jesus said. Consider also the interrelationship involved between citizens of the United Stated and the Bill of Rights guaranteed by our Constitution.

Chapter Six: The Hope Factor

Where Hope Comes In

As previously explained, there are forces employed in creation that have no physical component but are yet capable of acting upon physical objects such as the force of gravitational attraction that emanates from the sun to hold the earth in orbit. There is also the force known as "electrical energy" that is produced by the proton/electron relationship. The existence of these forces is discernable because they act upon physical objects that can be observed and monitored by the sensing mechanisms possessed by human beings. There are also forces with no physical component that act upon and affect other forces with no physical component but which can be observed as producing a change in behavior. The later, that take place within the mental faculties, are commonly called psychological. It is not the purpose of this discussion to define how such forces work but to demonstrate that such phenomenon does work.

In previous discussion it was observed that the term "faith" was appropriately defined as "trusting in a belief" and the term "love" was appropriately defined as "the willing of the good of another". In those discussions it was pointed out that in faith it is the terms and conditions of the covenant of the faith contract that determined how behavior is affected. It is the content of one's belief that determines behavior. When one believes in God, behavior will be influenced to behave to please God, which constitutes love of God. In this regard, love is the product of a belief. These are examples of forces with no physical component acting upon and affecting other forces that also have no physical component but which forces produce an observable change in behavior. Hope falls into that same category as faith and love.

The History of the Term Hope.

As with faith and love, hope came into existence along with the greater intellect and freewill that was endowed in humans. It is interesting to note that while not discussed or specifically addressed, hope has been functioning since the time of Adam and Eve. Surely Cain must have hoped that the Lord would be pleased with his offering since he was dejected when the Lord was not pleased. Abraham must have hoped for a son with Sarah. It is especially interesting to note that the term hope or inference of that term does not appear in the Gospels especially since it is the hope of salvation that is at the heart of the ministry of Jesus. As with the term faith, it was Apostle Paul who began using the term in his letters to the Christian churches that he supported. He told the Romans in 8:24, "Now that we are saved, we eagerly look forward to freedom. For if you already have something, you don't need to hope for it." Again in 15:13 he wrote, "So I pray that God, who gives you hope, will keep you happy and full of peace as you believe in him. May you overflow with hope through the power of the Holy Spirit." In 1 Corinthians 13:13 Paul said, "There are three things that will endure—faith, hope, and love—and the greatest of these is love."

It just seems odd that since Apostle Paul gave such importance to hope and believed it was something to last forever and since it is the cornerstone of the covenant given by Jesus, why that term did not receive more attention at that time. Perhaps because it was accepted as natural to hope and therefore was not recognized as needing emphasis or nurturing.

The Role of Hope

Definition of the term hope is generally accepted as being an expectation that a desire will be realized. A desire can be for a good end or an evil end. Some people might hope for salvation while some might hope that a misfortune be brought onto someone. But the term hope has a deeper meaning. Apostle Paul tied the term hope into his definition of faith in Hebrews 11:1 where he concluded that faith is: "—the confident assurance that what we hope for is going to happen."

The Holy Bible, New Living Translation, Tyndale House Publishing, defines hope as "confident trust with the expectation of fulfillment."

There can be little doubt that hope is in the category of a belief and therefore is an integral element in the faith process where the conscience influences freewill when engaged in behavior decisions. Hope is based upon faith, which as explained, is produced by trust in a belief. Hope is in the same category as prayer. One must believe that a particular matter that is hoped for, or prayed for, will happen. From that perspective it is seen that even though having no physical component, hope and prayer are motivators of behavior since they set a frame of mind or an attitude. Hope makes life worth living when it is directing behavior in the pursuit of desired objectives, even when there are unpleasant experiences that must be overcome in order to realize those objectives.

While hope and prayer are essentially similar, they are the same only in that they both are motivators of behavior. Hope is essentially a prayer to ones own self that produces motivation to pursue the objective of the hope. Prayer, while directed to God also has the effect of producing a frame of mind that influences the conscience that in turn influences behavior. Hope can be strengthened through properly directed prayer since by that act the frame of mind is moved to stimulate behavior that will produce the strengthening of hope. Hope can also be strengthened through prayer when accompanied by faith in the Holy Spirit.

The strength of hope varies depending upon the degree of the faith upon which that hope is based. Hope can be adversely affected by low self-esteem (faith in self) or by faith weakened by lack of confidence (trust) in the beliefs upon which faith in self is based. Anything that erodes the degree of faith also degrades the strength of hope and nothing erodes the degree of faith more than uncertainty regarding the truth of the core beliefs upon which faith and hope are built. That is why uncertainty over the creation process, the first truth, such as is produced by the ongoing creation versus evolution argument, is so destructive.

Chapter Seven: Bad versus Evil

Clarification.

There is a tendency for people to confuse the term bad with the term evil. In fact, even the concordance/dictionary in the Holly Bible, New Living Translation[23] defines evil as: "bad, sinful or morally reprehensible, of the Devil; something that brings sorrow, distress, or misfortune". Such definition leads to confusion. Frequently we hear people lament following a misfortune or a catastrophic event on how could God allow such a thing to happen, even to the extent of saying that such an event proves there is no God. For some, differentiating between bad and evil never enters their mind since they have been lead to believe that God is responsible for every event that occurs, especially those that result in misfortune that culminates in sorrowful experiences. That type of reaction results from a failure to differentiate between bad and evil. Bad is not a synonym for evil. Bad is a variation of good. There is no good in an evil act, just variation in the degree of evil, but evil acts often result in bad experiences. The built-in behavior characteristics of the elemental energy and particles used in the creation process do not produce evil. Animals actions, accidents or natural calamities are not of themselves evil. Only a human exercising freewill can produce evil.

There is also a tendency to think of evil acts in the context of Adolph Hitler or Joseph Stalin or something that is morally reprehensible, sinful, wicked or produced by a cosmic evil force. Evil is a deliberate intentional behavior resulting from thoughts formed in the mind of man by freewill reasoning whether by Hitler or Stalin or anyone else. Evil behavior is not an act taken against religious law but is behavior absent of consideration for the good of anyone other than the perpetrator. Evil behavior results in the harm of others. Jesus defined evil behavior as murder, adultery, theft, lying, slander, and sexual immorality (which was left undefined). These are behavior that, in addition to being sinful, disrupt peaceful co-

[23] Tyndale House Publishers, Inc © 1996

existence among people so are the type of behavior handled by civil authority. The classification of these acts as evil forms "core beliefs" that must be recognized before a viable society can exist. A problem arises when through the application of Relativism the definition of evil producing acts are liberalized to the detriment of the "law and order" that forms the basis of a viable society, the goal of the Creator God[24].

The Source of Evil

As with the term faith, the Scripture records no attempt to explain why there is evil or what causes evil behavior until the advent of Jesus. In Matthew 15:19, Jesus said: "For from the heart comes evil thoughts, murder, adultery, all other sexual immorality, theft, lying, and slander." In Mark 7:20-23 that explanation was expanded where it is reported that Jesus said: "It is the thought life that defiles you. For from within, out of a person's heart, come evil thoughts, sexual immorality, theft, murder, adultery, greed, wickedness, deceit, foolishness, envy, slander, pride, and eagerness for lustful pleasure. All these vile things come from within; they are what defile you and make you unacceptable to God." Luke 6:45 records Jesus as saying: "A good person produces good deeds from a good heart, and an evil person produces evil deeds from an evil heart. Whatever is in your heart determines what you say."

Apostle Paul addressed the matter in a manner that shows that he recognized the impact that belief in God and freewill has on evil behavior. In Romans 1:21 he wrote: "Yes, they knew God, but they wouldn't worship him as God or even give him thanks." In verses 28 through 31 of that chapter he wrote: "When they refused to acknowledge God, he abandoned them to their evil minds and let them do things that should never be done. Their lives became full of every kind of wickedness, sin, greed, hate, envy, murder, fighting, deception, malicious behavior, and gossip. They are backstabbers, haters of God, insolent, proud, and boastful." From these words it is clear that Apostle Paul believed that people's hearts or conscience

---

[24] Isaiah 9:7

were freed to produce evil when they refused to acknowledge and worship God. In Galatians 5:16-17, Paul wrote: "—live according to your new life in the Holy Spirit. Then you won't be doing what your sinful nature craves. The old sinful nature loves to do evil—." And in verse 19 he wrote: "When you follow the desires of your sinful nature, your lives will produce these evil results—." In Ephesians 2:2, Paul made a profound connection when he wrote: "You used to live just like the rest of the world, full of sin, obeying Satan—. He is the spirit at work in the hearts of those who refuse to obey God." Correlation and summation of these biblical passages credits freewill reasoning responding to Satan rather than belief in God as the influence that allows evil producing thoughts.

Not Who But What.

There is an important fact that requires reiteration at this point. Faith relationships are established when individuals trust in a belief. That phenomena works regardless of the content of the belief being trusted as expressed by Apostle Paul in Romans 6:16 where he wrote: "Don't you realize that whatever you choose to obey becomes your master." Paul's conclusion follows the statement of Jesus recorded in Luke 6:45 where he said: "Whatever is in your heart determines what you say." For example, it is a mistake for people to believe that their faith in the God of Abraham has the same meaning and will produce the same behavior for Hebrews, Christians, and Muslims when the content of what they believe about God is not the same. The most significant difference is that the content of the Christian belief includes Jesus the Son of God, the Word of God that Jesus delivered in behalf of the Father, and The Holy Spirit. It is also important to recognize that Jesus did not come to abolish the Law of Moses or the writings of the prophets but he came to show how to fulfill those laws by following the new commandment to love the Lord your God and your neighbor as yourself. That constitutes a significant variance with the Muslim faith adopted by the descendants of Ishmael and the Hebrew faith where belief content, not significantly unlike the belief content adapted by the Muslims, does not embrace the Word of God revealed by Jesus the Son of God. Since it is belief content that influences freewill behavior decisions, the impact of the differences in belief content of these faiths is reflected in the behavior of the followers of these religions beliefs.

As previously explained, freewill responds to what is believed not in whom belief is professed. A professed belief in God for example is meaningless since the content of such belief is subject to variation as stated, but especially when the meaning of words that defines the content of the belief are altered or ignored through the application of Relativism. For example, there are a considerable number of Christian Faith Denominations that profess their faith based upon the Nicene Creed yet the content of their belief deviates to such an extent that it is hard to believe that they profess the same faith. Beyond that, there are individuals who profess based upon the Nicene Creed yet actively practice Relativism in the application of the Word of God to their behavior. Classic examples are individuals and crime families such as the Mafia who proudly and publicly display their religious affiliation yet actively engage in acts clearly defined as evil by the Word of God that should provide definition to their profession of faith.

Good Out Of Evil

Humans, whether they believe in God or not, possess the intellect sufficiently powerful to reason and thereby direct behavior in response to their reasoning. Since reasoning can be influenced by experiences it is often stated that some good derives from bad or evil behavior since we see and dislike the experiences of certain behavior. That is a fact and is a factor in people being drawn to a source of good when the better light emitted by godly people would be recognized as an example to be emulated. It works in reverse, as exemplified by the World Trade Center destruction, to cause people to shun or avoid those beliefs that condone evil pursuits. It also has been shown as fact that those who have established a personal faith relationship with God, who have love of God in their conscience, are less likely to engage in evil behavior. Conversely, it has been shown as fact that those who have not established a deep faith relationship based upon obeying God are more likely to engage in evil behavior, as though more prone to being influenced by what is called Satan or the Devil.

Enter Relativism

As used in this instance, Relativism is defined as the practice of applying definition to a word or situation based upon a perceived relevance that produces a response more favorable to the perceiver than would result otherwise. A classic example of Relativism (some call it Liberalism) came out of a recent well publicized hearing[25] where in response to a question, the response was conditioned with a declaration that "it depends on what the meaning of the word is, is." In this example the application of Relativism impacted both legal and moral issues, which is often the case since a "law" is usually involved and it is the purpose of Relativism to change the meaning of a law in the perceiver's mind, whether secular or religious. This is not a new practice and has become the modus operandi of the trial lawyer profession and in the method employed by political parties and others with a liberal agenda, especially when dealing with contract or agreement specifications. The first reported application of Relativism occurred in the Garden of Eden when Eve succumbed to the relativistic argument of Satan who convinced Eve that God didn't really mean she would die if she ate the fruit. It shows again in three instances when Satan tried to use Relativism in tempting Jesus during his forty days and forty nights in the wilderness[26].

Relativism was given new direction following the publication of Darwin's Theory of Evolution and Einstein's Theory of Relativity. It gained further impetus during the Great Depression/New Deal Era, World War II and hippy age experiences of the 60's. Relativism has become rampant through the subtle manipulation and alteration of the definition of key words that then change the meaning of key requirements of law or contract terms such as the content of the beliefs that form the basis of a faith relationship. By alteration of the meaning or interpretation of the terms that form the basis of a belief,

---

[25] Testimony of president Bill Clinton in the case involving Paula Jones
[26] Matthew 4:1-10

deviation from the intended meaning can be practiced without engaging the individual's conscience. For example, in Mark 9:20-23, Jesus is recorded to have said; "It is the thought life that defiles you. For from within, out of a person's heart, come evil thoughts, sexual immorality, theft, murder, adultery, greed, wickedness, deceit, foolishness, envy, slander, pride, and eagerness for lustful pleasure. All of these vile things come from within; they are what defile you and make you unacceptable to God." Those are the type of key words that are the target of Relativism. When the meaning of those words are changed the meaning of a persons profession of belief in God or Jesus or of any other faith contract for that matter, would be altered, as would the basis used in the freewill reasoning process. The words stay the same but the meaning of those words is changed and thereby the belief held in the conscience. The application of Relativism is the principal factor responsible for the differences in behavior that exists among people who call themselves Christians and make the same Nicene Creed Profession of Faith or Muslims with a varied interpretation of the Koran. In the face of Relativism, a basis of faith such as the Nicene Creed, the Bible or the Koran, are made meaningless.

In another example, a major division and deviation in belief was produced by a relativistic interpretation of what Apostle Paul meant by his statement that salvation is achieved by faith alone. As previously explain Apostle Paul recognized that faith in God influences people to obey God's will, as did Abraham. Therefore, it follows that those who obey God will through faith love God and that love will be shown by behavior producing good deeds. The Apostle James in his letter to the Jewish Christians, chapter 2:18 said the same thing in another way: "I can't see your faith if you don't have good deeds, but I will show you my faith through my good deeds".

In the political and secular judicial arenas, the application of Relativism is rampant. A recent well-publicized example occurred in the political/judicial world of politics when on October 3, 2002, the New Jersey Supreme Court ruled to approve a deviation from a precisely written law concerning the conduct of elections. In this instance, a variance of a stipulation in the law was based upon a contrived relevance that the replacement of a candidate who voluntarily withdrew from a statewide contest after the date allowed by the law, would deny the voters an opportunity to vote for a candidate representing the political party of the candidate who withdrew. The voluntary withdrawal in this instance was precipitated when evidence showed that a replacement candidate would have a better chance of winning than the candidate who withdrew. This convoluted relevance finding of the judges allowed a variance in the meaning of the words in the law without changing the wording of the law and without engaging their conscience. In this case it was the Justices of the State Supreme Court who engaged in the practice of Relativism. It has become common practice for judges to impose their own personnel agenda's by altering the meaning of words to alter the intent of laws and thereby by pass the legislative processes.

One of the most detrimental of all applications of Relativism involves the subject of creation itself. It has become the practice by certain influential segments of our society to deliberately and purposefully deny that our universe is the work of a creator while publishing misleading theories rather than letting the facts concerning the process of creation speak for themselves. With the role of the Creator, the first truth, the basis of all truth, decimated in the minds of many, then the conscience of those many are void of the one belief fundamental to the functioning of built-in behavior control and in combating the influence of Satan.

There is no way of knowing before hand what impact Relativism has on the objective of gaining life after death but judging by the growth of immorality in our society, the behavior of people is being adversely impacted by something. It is also evident that when the practice of Relativism is accepted without objection, the practice of Relativism becomes accepted as normal, which promotes further such practice including denial of a creator.

Dealing With Bad Experiences

Experiences with bad results occur. Experiences with good results occur, as do experiences with results somewhere between those extremes. Experiences with bad results occur as the result of accident, as the result of evil intent, due to natural calamity or due to unintended mutation of the composition and arrangement of the elemental particles and energy of the assemblies of which things are made, especially those involved in producing human beings. In all instances, except for those produced by evil intent, whether or not an individual is involved, bad experiences are essentially a matter of luck or bad judgment by individuals and not the result of actions or inactions on the part of God. As previously explained, God does not act directly but acts through faith contracts with individuals as was the case with Moses.

While there is no known guaranteed protection from bad experiences, there are certain behaviors that enhance the risk of their occurrence. The consumption of mind-altering chemicals produces a high risk factor as does engaging in activities that involves trusting in the behavior of man made creations such as automobiles, planes and buildings. Mutation, the altering of the designed configuration of key critical assemblies of particles usually DNA, results primarily through the consumption of certain chemical substances or in the course of the marrying of maternal and paternal DNA contributions in the reproduction process. Many of the bad experiences resulting from natural calamity occur when people persist in taking residence in locations that are historically prone to the occurrence of natural disasters.

While the forces that produce natural calamities, often categorized as acts of God, are produced by circumstances directly attributable to the functioning of the creation process without the intervention of mankind, such experiences generally do not cause people to react by questioning the motives or non-existence of God. It is as though such acts affirm the existence of a Creator who is not all-powerful and perfect without fail and when those events occur, there is nothing anyone can do to prevent their occurrence. On the other hand, the questioning of God for bad experiences that were not of God's doing may be by people who are wondering why they were punished when others who engage in comparable behavior are not punished. When faced with bad experiences it has been demonstrated that those who have a well-established faith contract with God are able to better cope with those type experiences.

Understanding the Role of God.

As has been previously explained, creation of the universe including humankind involves the assembly of elemental particles possessing precise eternal physical and behavioral properties and forces that remain connected to the source of all energy at the center of the universe by the energy with no physical component that scientists call affinity. Even though the elemental particles and the forces employed in assembly remain connected with the source of all energy from whence they originated, each component stands alone with specific unique precise eternal behavioral characteristics. When assembled in building block fashion to produce more sophisticated assemblies, those assemblies with components remaining connected to the source are in turn stand alone with new specific unique precise eternal behavioral characteristics. That scenario follows through each progressively more sophisticated assembly up to and including human beings. It is important to recognize that each elemental particle and each force applied in assembly and all succeeding assemblies function according to the precise behavioral characteristics built-in. The manner in which creation was to function was built-in from the outset, even the "mechanism" for freewill. Considering the time and the extent of progressive evolution that has been involved since creation along with the magnitude of the creation effort, it must be conceded that the creator did an outstanding job of envisioning and providing for subsequent events. It is right that we should give the creator thanks and praise. However, it is still possible that something might go array whether by accident or because all possibilities could not be accounted for when the behavior characteristics of the building blocks were set. Some functions such as earthquakes and the weather are not linked directly to the behavior of the building blocks.

Except for the energy seen as producing gravity and proton/electron forces, and for the power of the Holy Spirit, we are here on earth alone with our fellow man. That is why the establishment of a faith relationship that produces love of God and love of fellow man is so important. It is through that faith relationship that hope prevails even in the face of bad experiences.

Anti-Freedom Forces.

Whether or not deliberately and purposefully subtle and whether or not out of ignorance or knowingly directed, there are forces at work that subvert the freedom to exercise freewill. Examples include the Marxist philosophical beliefs that became the foundation for Communistic governments and other religion based philosophies (Marxist philosophies are religion based) that are being applied to set the foundation of civil governments, especially when those philosophies restrict and stifle the freedom to practice other religious beliefs. Such stifling and restrictions have usually been applied under cult, Theocratic, dictator or monarchy type governments.

But, there are other philosophies that stifle freedom including those whose proponents are either oblivious out of ignorance of the nature of their intentions, or who resort to all manner of subterfuge to conceal their intentions. It is usual for those proponents to champion causes that have motives that are on the surface defensible as promoting various freedoms. These include such causes as atheism, animal rights, civil rights, women's rights, protection of the environment and similar other extreme liberal or extreme conservative agendas that collectively function in the same manner as does Relativism and the evil producing philosophies of Satan. Those holding such agendas work to stifle and restrict the freedom to exercise freewill.

The proponents of anti-freedom force agendas have one thing in common. They are unwilling and/or unable to allow the built-in freewill mechanism to operate without restriction as designed. Their ideologies and philosophies, whether religious or civil, do not or cannot compete in an environment without the imposition of restrictions on the exercise of freewill. The ability for those forces to succeed requires the stifling of freedom of others and either all such agendas have religious connotations or all such agendas do not have religious connotations since all involve the same behavioral control mechanism.

The preceding argument is made with but one stipulation. The civil governing authorities of the various individual societies have the right and responsibility to outlaw or make illegal various activities or behavior that are deemed intolerable due to their adverse impact on their society since, unlike the Kingdom of God, enforcement cannot be deferred until judgment day.

Chapter Eight: Immoral Sexual Behavior

Defining Immoral.

Dictionaries do not provide definition for the word immoral except to say it means "contrary to established moral principles." To that end, Merriam-Webster[27] for example, defines the term moral as: a. of or pertaining to principals of right and wrong behavior, or: c. conforming to a standard of right behavior, or: d. sanctioned by or operative on one's conscience or ethical behavior. The dictionary/concordance of the Holy Bible, New Living Translation, defines immoral or immorality as: "behavior conflicting with traditional (biblical) moral values". Those are relativistic definitions if there ever was such a thing!

References within the Scripture are equally vague when it comes to defining the meaning of the term immoral. The terms immoral or immorality were not used until the advent of Jesus. The terms sin, wickedness, righteous or unrighteous were used in the Old Testament when referencing behavior. In Genesis 6: 5 it was the peoples wickedness and in 6:9 it says Noah was a righteous man. Again in Genesis 15:6 it says that the Lord declared Abram righteous and in Genesis 18:20 it says that the people of Sodom and Gomorrah were extremely evil and wicked. In Exodus 32:7 it is said that the Lord told Moses that the people he had brought from Egypt had defiled themselves. In Leviticus 4:1 the term sin was used to describe those who do anything forbidden by the Lord's commandments. One of the first places where the term immoral was used was in Proverbs 5:3 in speaking of the lips of an immoral woman.

The most surprising revelation is that the Catechism of the Catholic Church[28] does not mention morality or immorality in those terms. Perhaps it is true that it depends upon the meaning of what the word is, is.

---

[27] © 2001 by Merriam-Webster, Incorporated

[28] English Translation © 1994, United States Catholic Conference, Inc.—Libreria Editrice Vaticana

Even though "immoral sexual behavior" apparently constituted a major problem throughout history from the time of Adam, it is of interest to note that Jesus did not send out his disciples to save people from sin or immorality but to announce that the Kingdom of Heaven is near. The concern of Jesus was to heal the sick and cast out demons. The message of Jesus[29] was that "people are defiled by an evil heart from which come evil thoughts, murder, adultery, all other sexual immorality, theft, lying, and slander" where the cure was belief in him. It is of further interest to note how Apostle Paul addressed the matter of immorality in Galatians 5:19—20 where it is said: "When you follow the desires of your sinful nature, your lives will produce these evil results; sexual immorality, impure thoughts, eagerness for lustful pleasure, idolatry - - -", along with others. For some reason, both Jesus and Paul singled out sexual behavior but not adultery or the many other types of evil behavior, to have an immoral aspect. Clearly, for Jesus the way to combat the problem was to get people to add the love of God to their conscious so as to produce trust in their belief of him, which constitutes faith. The Scripture gives no indication of any thought of restricting freewill or of imposing any other restrictions as a means of combating this apparent serious continuing immoral behavior problem. Rather, the actions taken by Jesus affirm God's continued faith in the behavioral control system that is inherent in the human anatomy as previously defined. The principal action taken by Jesus in that regard was the encompassing of the Laws of Moses into two simple commandments[30] by which freewill was unleashed and let go free.

Defining Sexuality.

Merriam-Webster[31] defines sexuality as: "the quality or state of being sexual: **a:** the condition of having sex **b:** sexual activity **c:** expression of sexual receptivity or interest especially when excessive." This appears to reflect the best of the currently expressed

---

[29] Matthew 15:19

[30] Matthew 22:37-40

[31] © 2001 by Merriam-Webster, Incorporated.

definitions although not of much help in defining "sexual immorality" as expressed in the Scripture.

Defining Sexual Immorality.

There is little to be gained by attempting to sort out the intent of the Scripture in this regard based upon the stated definitions for sexuality and immorality. The Catechism of the Catholic Church goes into considerable explanation of this matter under the subject of Chastity. Perhaps the most definitive explanations are provided in Leviticus 18 and Romans 1:20-27. Anyway, it is not the purpose of this work to develop a definition for that expression but rather to point out that the definitions presently available are totally incapable of clarifying the intention of the Scripture or even civil law. For the purpose of this work it is sufficient to recognize that some kind of sexual activity involving human beings beyond that of adult married heterosexual couples, has been producing a serious human behavioral problem from the beginning. It is not what specific behavior constitutes the problem but why the method of behavior control endowed in humans has not been working, as it apparently should in the first place, especially since even avowed religious ministers and pastors are not immune.

A Review of How Sexuality Works.

Sexuality is first and foremost an inherited instinct, typical of those instincts endowed to all mammals by the Creator for the purpose of perpetuating life. Sexuality had a purpose even before the creation of humans with their superior intellect. Sexuality is beyond a doubt, intertwined with personality characteristics that are also inherited. Since gender is involved in establishing personality characteristics, gender is also a function in defining sexuality characteristics. That is, in the animal kingdom, males with their more aggressive personality characteristics are generally more sexually aggressive while females are by nature generally more sexually seductive. Generally speaking, females "turn-on" the interest of the male to insure that interest is not lacking in their choice of a mate or their offspring, which may be as a type or personality characteristic rather than a specific appearance. That is not to say that the instinct for sexual activity is more or less powerful in either gender. Both male and female instincts vary in accordance with the inherited instinct characteristics but in human beings the power of those instincts stand to be multiplied through the action of superior intellect acting in response to freewill. With their superior intellect, humans can think, reason, imagine and lust. Even though faced with that formidable influence, control over sexual behavior still relies solely upon the influence of beliefs embedded in the conscience that are applied through the application of freewill reasoning. Since the characteristic of inherited sexuality varies between individuals, it is apparent that the strength of conscience embedded beliefs must likewise be adjusted to compensate if behavior is to be consistent between individuals. That is, some individuals need greater attention to insure that proper beliefs have been established.

However, as if behavior control under the influence of the sexuality and personality instincts did not present problem enough, the sexuality instinct carries a "double whammy." It is the characteristic of the instinctively driven sexual activity to produce a very "rewarding" sensation of pleasure in the involved parties. By that reaction, each of the parties is moved to "willing the good of" the other which instills an additional pleasure from the feeling of being loved that adds to the sensation of the pleasure experienced. It is the uniting of those pleasures that makes relations in a monogamous marriage so rewarding and which makes control of such activity through free will reasoning especially challenging for those outside of monogamous marriages. By conventional standards one might say that the deck is stacked. That is especially true in the case of adolescents where the power and repercussion of those influences are not fully appreciated, but upon all people where the countering beliefs have not been developed sufficiently to maintain control over those "double whammy" instinct driven activities. Understandably, those who have not built a conscience belief system based upon trusting in the beliefs that are produced by the love of God will be at the mercy of Satan and Relativism on the one hand and reasoned fears on the other.

Fear, The Unheralded Deterrent

The heart of the behavior control system endowed in human beings by the Creator is a conscience wherein control of behavior is produced by the influence of trust in impressed beliefs. The Creator God, especially in the Judeo/Christian/Muslim religions, relies upon the love of God to impress resulting beliefs in the conscience of individuals. On the surface, all seems to hinge upon love or as St. Thomas Aquinas would say, willing good. The Relativistic movement has locked on to the love aspect of the beliefs taught by Jesus to down play the role of fear as a means of influencing behavior. In fact, there have been and still are many with liberal leanings who advocate that fear is detrimental and is not appropriate for use in behavior control. Let it be recognized that fear is one of if not the most powerful of the motivators for influencing the freewill direction of behavior. It could be argued that fear is not truly a motivator but that it is the alarm that sounds when the conscience detects a challenge to an impressed belief. When viewed in that light it appears that fear is related to faith. Consider that while a tightrope walker retains faith in the ability to walk the rope safely, fear is suppressed or turned off and the walker is confident. When the faith of the walker is not strong, the walker is not confident and fear warns against that action. From this perspective it is apparent that fear requires the existence of a belief before it can work.

While not emphasized, fear is an important element in the word revealed by Jesus. For example Jesus repeatedly charged that unless people believe in him and obey they could never enter the Kingdom of Heaven or threats to that effect, which rely upon fear for effect. In Luke 12:5, Jesus is reported to have said: "But I'll tell you whom to fear. Fear God, who has the power to kill people and then throw them into hell". The fear of eternal damnation motivates people to attempt to strengthen their faith in God.

The fear of not gaining eternal life is but one example of how fear motivates. The fear of loosing favor of individuals or peer groups has a very powerful influence on behavior especially with young people. In a similar manner the fear of punishment or reprisal or even of harm from hazardous situations, not the least of which is the fear that engaging in certain behavior may result in the contracting of sicknesses or unpleasant experiences, will influence behavior.

The Rebellious Youth Phenomenon

This is a well-known and often studied phenomenon that takes place even in the face of the best of religious teaching, even with literacy blessed quality parental preparation and supervision. Rebellious behavior is often attributed to individuals wanting to test their faculties to see if they have what they believe it takes to be an adult. Their deduction is probably correct and they in most cases probably do have the faculties. That leaves the question, why do things go so wrong at that time of transition in their lives that occurs during adolescence? There is no question that young adults generally have not had enough experiences to establish a proven reliable conscience and some may not have established a sound faith relationship with God. Then the question should be, what can be done to prepare the conscience of young people so that they will be less likely to engage in destructive behavior? It is well known that understanding cannot be forced into an unwilling or non-receptive subject. We have all heard a story about the preacher's son or daughter. If teaching, preaching, threatening, and supervision do not control the rebellion phenomenon, could it be that the conscience of our youth is being corrupted by the insertion of defective beliefs?

As previously discussed, love of God the creator is the key to the development of a properly functioning freewill driven behavior control system regardless of how that love relationship is established. It was previously shown that faith produced by trusting in a belief where God the creator is responsible for our existence (the first truth) is the first requisite to the development of love of God which is the key factor in behavior control. When we allow our children to be taught otherwise, that first requisite is missing and the basis for love of God will not have been established and the conscience will have been prepared to embrace Relativism. Perhaps a worse scenario develops when a child is taught at home or in church, synagogue or mosque that God is responsible for creation but then taught in school by those who have been entrusted to educate our children, that anyone with brains knows that there is no creator God. That essentially is what our children, who have not had experience necessary for the development of a strong faith relationship with God, are being forced to face in today's society. What beliefs will those adolescents register in their conscience? What impact will that have on the freewill behavior choices they are required to make? Many faced with this contradiction rebel against their parental religious teaching even to the point of accusing their parents of "ruining" their lives. On such occasion it becomes easier to embrace the belief there is no Creator rather than to hold faith in the belief that there is a Creator, and that produces a flaw in their conscience that encourages Relativism.

A worse scenario results when the later example is reinforced through exposure to Relativism in action. When young adults see older adults get away with violation of moral beliefs and civil laws through the exercise of Relativism, they are caused to believe that the observed behavior is OK aggravating any flaw in their conscience. Exposure to relativistic behavior is especially damaging when the violation is by a high visibility person who occupies a high ranked position and/or who is in a role model capacity. Consider the case of Relativism where the President of the United States[32] proclaims that oral sex is not really sex under civil law, not to mention moral law. Consider also that the news is filled with other examples where Relativism engaged by highly regarded celebrities and corporate executives has been carried to such lengths as to eventually bring criminal charges. Who is there that would not recognize the potential for adverse impact when the formative consciences of young adults are exposed to such behavior? It is when such behavior becomes implanted in the conscience of individuals as acceptable, even to the point that it becomes believed to be a prerequisite to success, that the impact on the conscience of individuals and thereby the morality of a society should be obvious.

Evidence of the corruption of the behavior controlling system is seen by the incidences of immoral and evil behavior but the disturbing factor is the absence of remorse or shame or guilt or fear of

[32] President Bill Clinton

condemnation, by those who engage in such behavior. That by itself gives evidence to support the conclusion that the underlying fault corrupting the behavior system embodied in the design of human beings is serious. The ultimate example of corrupted conscience is seen in the number of instances where young adults resort to evil acts for the purpose of satisfying there urge for notoriety and fame such as the Columbine Colorado school shootings or the Muslim suicide bombers. These are hope-deprived souls where the lure for personal notoriety and fame has come to dominate their behavior controlling system.

Is there a solution? Is there some one thing that will prevent such behavior? Possibly not but there is something that might help bring the problem under control if it were just tried. It is possible; it is worth trying to do what can be done to insure that our educators, scientists and religious stop lying to our children about creation and the role of God in the creation process. Its time to stop hiding the truth behind unproven theories, myths, dogma, and traditions that have been made to suit the agenda of the theorists and which fail when put to the test. It is time to acknowledge that the nature of the Creator God and the processes of creation are revealed through the study of that which has been created. When those acknowledgements are made the truths that instill beliefs necessary for appropriate behavior will be revealed. Then our children will not be sent on life's journey on ships with out a rudder or the mechanism needed to navigate treacherous waters without hope. No tunnel without a light at the end. No reason not to understand the purpose of life and the process for fulfilling that purpose.

Remember! This is not simply a religious matter. The behavior of all things, whether in response to secular or religious intentions, begins with the first truth. If the first truth is denied, the truth upon which all truth depends, then there is no truth and if there is no truth there is no order to behavior. Remember too, we develop from a single cell that has no brain wherein knowledge and beliefs can be passed along. All of the knowledge needed to make behavioral decisions that are involved in navigating life's treacherous waters, must be learned. That leaves precious little time for adolescents to sort through what they have been taught so as to develop a dependable belief system, especially when what they have been taught is defective or worse yet, purposefully erroneous.

Chapter Nine: Freedom's Highway

Highways For Living

Any road will suffice when one takes on a journey where the destination is not defined. Under those circumstances, the objective is to travel down a road without concern for the destination but just concentrating on enjoying the ride. Even when a specific destination has been established, that objective becomes less inviting when the route to that destination is unclear or the road traveled is circuitous with numerous branch roads, or the route markings are confusing, or when faced with distractions, obstacles and congestion. This is the scenario that existed in the United States in the late 1950's when under the administration of President Dwight Eisenhower an Interstate Highway System program was implemented so as to straighten the way. The new interstate highway system is a work that has been in progress for decades so as to remove the hassle, difficulty and uncertainty of traveling thereby providing greater assurance that ones destination is easier to reach.

So it is too with the "highways for living" that are established to guide and facilitate the journey of life. These are "highways" that exist (and have existed) in the form of both secular and religious **belief** systems that establish and define the rules and provide direction for life's journey. Those belief systems and their "highway definitions" exist whether as Socialism, Communism, Monarchy, Democracy, Hinduism, Buddhism, Judaism, Christianity or Islamism. At the time of John the Baptist, the belief system of Judaism, implemented through Moses to guide the Israelites on their journey to the "Promised Land", had become a jumbled mess. The message of John the Baptist made reference to the prophecy of Isaiah[33] some 500 years earlier when he said: "There is a voice shouting in the wilderness. Prepare a pathway for the Lord's coming! Make a straight road for him—Then the glory of the Lord will be revealed and all people will see it together." The belief system implemented by

[33] Isaiah 40:3—5

Jesus to straighten the way leading to the destination called "salvation or eternal life", by drawing people to love the Lord our God as a means of directing traffic on life's highway, like the U. S. Interstate Highway System, is a work in progress. Also, as with the interstate highway system, the straight way implemented by Jesus does not work until one enters and begins traveling on the straight way and the traveler must remain alert and not become distracted or make a wrong exit.

Finding The On-Ramp

To take advantage of a straight way it is first necessary to recognize that the straight way exists. Then the next hard part of the journey is getting people to the on-ramp. On the way to the on-ramp there are many old highways, cross streets, distractions and obstacles that must be navigated and the proper way chosen. So it is with the straight way implemented by Jesus.

One of the problems that has to be overcome in the process of building a new straightway are the objections raised by those who have interests dependent upon keeping the old ways open for use since by their perception, their livelihood and importance is dependent upon the old way. To those people, maintaining the old way is more important than helping people to reach their destination. On the other hand, there are other old ways that do not have a defined destination since they keep going around and around hoping for reincarnation. Then there are the atheists who refuse to acknowledge that they are on a journey. They just let the driving to others without a care as to where the bus is going.

The biggest problem of all is to get those who administer the highway systems of the various religious orders to recognize that the objective before all else, as stated by the first of the commandments, is to bring people to love God. If that objective is not recognized by a religious order as their principle objective, such order should not be recognized as having a religious purpose and if that is the purpose of a religious order or even a necessary requisite to the success of a religious order, why should anyone object to use of the best straight way to reach that objective. For the love of God! That's the on-ramp, the love of God.

The Purpose of the Trip

The purpose for our existence is unquestionably the greatest of the outstanding mysteries. To find the answer it is necessary to begin consideration at the beginning when the first elemental particle of matter was created as was previously explained. When examined from that perspective it becomes self evident that there is a point, now at the center of the universe, that is the source of all energy and wisdom. There is one other key consideration; the wisdom of the creator (the truth) is in spirit form not directly accessible or understandable by human intellect. It follows then that the wisdom employed in creation would be revealed to human beings only through the examination, analysis and understanding of that which was created, whether in the form of the physical universe or as the inspired word of The Creator God. It follows further that an intellect capability mirroring that of the creator would be needed to ferret out an understanding of the wisdom (the truth) contained within the creator's work. Is that the purpose for human beings? Are we the mechanism created for the purpose of ferreting out the wisdom inherent in the creator's work and translating that wisdom into a form understandable by intelligent beings? You decide. At least that is a meaningful "makes sense" purpose, since regardless of what we may think and whether or not we agree, we exist as a "tool" of and at the pleasure of the creator and there is a penchant in each of us to understand our origin.

There is a second argument in that access to the wisdom of the Creator is necessary for humans to achieve dominion over the earth, which was God's first command[34]. Through living and dying, our lives contribute to the growing collection of knowledge just as the life and death of earlier life forms contributed to the progressive development of our environment. It is comforting to understand that there is a purpose for our trip even if we do not now know the

[34] Genesis 1: 28

eventual end product of the cumulative contribution of our travels. It makes sense that our trip would be more pleasant and productive when our intentions are directed through love of the creator our God and, for those traveling on the straight way, "global positioning" assistance provided through the Holy Spirit.

Pre-Abraham Roadways

The roadways of ancient times give indication that they were not designed and planned. The people of that time were in a sense pioneers in the journey of life for which there were no maps; no defining guidance. It must be assumed that their life's journeys were directed primarily through inherited instinct and experiences that produced learned understanding. Through superior reasoning, humans are able to differentiate between experiences that produce acceptable results and experiences that produce unpleasant results. Understandably, the extent of understanding then existing would have been minimal. However, subsequent studies have established that even then "wise men" such as Confucius existed who, possibly with divine guidance, developed beliefs that apparently guided behavior appropriate for circumstances then existing. At least mankind survived so there is nothing to be gained by finding fault with those belief systems except to point out that history records subsequent learned experiences and superseding divine guidance. There now exist "improved and better-engineered" highways that provide greater assurance of a more fruitful trip. If there are individuals who want to continue following those ancient roadways for life's journey when better straight ways are available, that should be the individual's choice since freedom to choose is our Creator endowed right even though God has subsequently provided better engineered highways. But, if God in fact is inspiring the directions of such systems, God should then be leading those people to the onramp of the better-engineered highway. Perhaps arguments intended to sell the merits of the newer better-engineered highway systems has been flawed.

The Post-Abraham Roadways

The Old Testament records the beliefs that defined the roadway followed by the Hebrew people. By that accounting a new highway system was initiated when Abraham was chosen to lead his descendents. This was to be a restricted highway limited to descendents of Abraham that had been marked by circumcision. Since the Lord himself initiated this roadway, the reason for excluding all people except those descendents of Isaac and Jacob is unclear. It is also mystifying why, as recorded in Genesis 21: 12-20, the development of this roadway had hardly been started before a branch in the roadway was established to accommodate Ishmael, the firstborn of Abraham born of Hagar, and the descendents of Ishmael. It is also important to recognize that both the Isaac-Moses branch and the Ishmael-Mohammed branch of the Abraham highway were established initially without rules to define the beliefs needed to direct travel. In the case of the Isaac-Moses branch, rules were not set until defined through Moses some 400 years after Abraham and then subsequently modified some 1,500 years later by the ministry of Jesus. In the case of the Ishmael branch, rules were not published until defined by Mohammed in what was called the Koran some 2,000 plus years following establishment of that roadway by Abraham.

The history of the problems the Israelites had following the Isaac-Moses highway is well documented in the Old Testament and was previously discussed in Chapter Three where an inherent defect was identified. That defect involved the absence of a method for the development of faith contracts between each individual Israelite and God which meant that individuals were not motivated to love God through freewill because his laws had not been written on the hearts and in their minds of each individual. Instead, the Isaac-Moses belief system that directed travel on that highway were mandated through laws that restricted freedom to apply freewill in making decisions. The directing laws involved covenants between Moses and later King David and God and not between individual Israelites and God. But that was not the only defect. As stated in the preceding discussion, all people who were not descended through Isaac, especially Ishmael and his descendents, were excluded from the use of that highway and thereby ostracized.

A branch in the Isaac-Moses highway developed with the advent of the ministry of Jesus although the apparent intent initially was only to modify the original Isaac-Moses highway system where The Word as written by Moses set the rules of the road. The ministry of Jesus followed hundreds of decades of decimation suffering and oppression during which time the efforts of innumerable prophets failed to convince the Israelites to abide by the rules of the road set through Moses. There upon, the prophets prophesied a change in the Isaac-Moses rules where it was said that God planned to make a new covenant that would "put my laws in their minds, and write them on their hearts[35]." Even though that prophecy came with a declaration that the new covenant "will not be like the one I made with their ancestors when I took them by the hand out of the land of Egypt", the new covenant subsequently advanced through Jesus has been rejected, even to this day, by those administering the Isaac-Moses highway. Since God brought the new covenant by revealing himself as a human rather than by speaking through a human prophet, as had been customary, understanding and acceptance of that modification of beliefs by the Israelites produced controversy and disbelief. Jesus as recorded in Matthew 10:34—35 predicted his ministry would produce such discord when he said: "Don't imagine that I came to bring peace to the world! No, I came to bring a sword." The modification God implemented through Jesus was drastic as though accomplished with one swipe of a sword and is recorded in Matthew 12:40 where, after announcing the two most important commandments, it is recorded that Jesus said: "All of the other commandments and all of the demands of the prophets are based upon these two commandments." With that one swipe of the sword God decreed the end of the theocratic Moses system of behavioral control and decreed activation of the built-in freewill system as previously discussed. In that manner, with those two commandments, God's laws were put into the mind and written on the heart, which is the conscience, of individuals who believe in him just as prophesied by the prophets of Israel. ***With***

[35] Especially Jeremiah Chapter 31: 31—34

those beliefs impressed in the conscious of those who believed, the ability to influence behavior of individuals was established without restricting freewill. It is important to make note of the fact that as declared in Matthew 5:17—19, this action by Jesus did not abolish the Law of Moses or the writings of the prophets. Instead, it was by this action, that faith contracts of believers were established between each individual and God rather than between Moses and God so that individuals would be influenced to fulfill the Law because they wanted to rather than because they were coerced to under duress by the threat of God's wrath.

The Ishmael-Mohammed Highway

The Ishmael-Mohammed highway floundered without direction from the time of Abraham to the time of Mohammed (570-632 CE) a period in the order of 2,500 years. In fact, by the time of Mohammed the Ishmael highway had been taken over by a multitude of pagan belief systems. Mecca, the site of the temple and the shrine Ka'ba, which is itself in effect an idol that tradition says was built by Abraham and Ishmael as was the custom of Abraham, had been made a ritual site of the pagan idol worshipers. Throughout that period, Ishmael and his descendents were ostracized by the Jews and following the advent of Jesus, ignored by the Christians even though there is evidence that the Judeo and Christian religions may have been practiced in Mecca to some extent during those intervening years.

Those were the circumstances prevailing when Mohammed set out to restore the Ishmael highway. In that process, Mohammed retained the Abraham-Ishmael connection, including the shrine Ka'ba, and its association with the Hebrew and Christian belief system. Mohammed concentrated upon eradicating the pagan idols and their worshipping practices and in their place compiled a new set of beliefs that mirror in many respects the beliefs documented by Moses plus some of the beliefs documented by the Gospels, ***but not the belief that Jesus was the Son of God, or that the most important commandment is to love God, or that salvation is the reward of belief in the "Good News" brought by Jesus.*** Rather than concentrating upon God's new commandment to love God and love your neighbor, which commandment is at the heart of the ministry of Jesus, Mohammed placed special emphasis upon his Moses like concept that there is no god but Allah; that he Mohammed, was the prophet of Allah; and the beliefs he documented are the only true beliefs. Since the god Allah of the Muslim belief system developed by Mohammed is inferred to be the God of Abraham, just as recognized by Judeo/Christian belief systems, that is a good place to begin developing an understanding of the Muslim faith system.

First of all, it needs to be recognized that Mohammed engaged in the restoration of the Ishmael highway and not the creation of a new highway, as is generally alleged, since the connection with Ishmael and Abraham and the shrine at Mecca were retained as the heart of the restored highway, as were some Abraham/Ishmael traditions and beliefs that were included in the Koran. The tasks faced by Mohammed when he endeavored to restore the Ishmael highway were similar in many ways to the problems faced by Moses when he took on the task of rescuing the Abraham-Isaac highway. It is important to recognize that at the time of Mohammed, the temple at Mecca had been taken over by a collection of idol worshiping barbarian pagans that had to be ousted by military style force and the descendents of Ishmael had to be convinced to understand that there is no god but Allah and not to engage in idol worship. In the process of confronting that problem, the instructions produced by Mohammed contained some strong assertions against the idol worshipping infidels of the day which assertions have been interpreted by some followers of Islam to include Judaists and Christians. In the interest of confronting idol worship, Mohammed emphasized what has become the Shahada or Creed of the Islam faith that states that: "there is no god but Allah and Mohammed is the prophet of Allah."

On the face of it, the establishment of a Creed that there is no god but Allah and Mohammed is the prophet of Allah, may not seem to present a problem. However, when compared with the first and greatest commandment that sets the philosophy of the Christian faith system, and in light of the process by which behavior in humans is influenced as previously explained, the basis of a problem begins to immerge. To help illuminate the problem, consider that the Christian first commandment and the commandments of Moses, exhorts each individual to love God and ties God to the patriarch Abraham. Of significance is the fact that neither the Judeo nor Christian first commandments make any effort to glorify the messengers Moses or Jesus. All honor and glory is directed to the living God so that there is no successor involved. By comparison, the Islamic commandments do not exhort the love of God; do not preface their documented beliefs with the acknowledgement that before all else, Allah is the creator of all; but they do mandate a belief that Mohammed is the only true prophet. In other words, only the belief system of Jesus that is based upon loving God the Father, Creator of all, before all else is in accord with the behavior control system built into human beings. The first commandment of Jesus exhorts each individual to "will the good of God" whereas; the Moses first commandment exhorts the Israelites as a group to believe that it was the God who rescued them from Egypt, the God of their ancestors Abraham, Isaac and Jacob, who was speaking to them where his first command was not to worship any other gods and not to make any idol. By contrast, the Islamic Creed demands that individuals believe that there is no other god but Allah; that Mohammed is the only true prophet of Allah; and that Islam is the only true belief system. Based upon the Islamic Creed, the god called Allah is not the Creator God of Abraham, Isaac and Jacob; and the Father of Jesus.

It is the last two beliefs mandated by the Islamic Creed that prohibits the followers of Islam from believing what Jesus taught while at the same time instilling hostility and hate against those who do no embrace Islam. ***It bears repeating***: only the belief system instigated through Jesus implores love of (willing the good of) God and fellowman as the instrument in individual personal faith contracts to influence the behavior of individuals. Only the faith system instigated through Jesus leaves State governing to "Caesars" and personal behavior under the freewill control of individual faith contracts.

The Theocracy Factor

It is of interest to note other similarities between the Koran and the Torah. Both forcefully condemn the worship of idols. Both embody instructions or laws for the express purpose of mandating personal and societal activities which, when followed produces a Theocracy type society wherein freedom to reason and to react in response to that reasoning (freewill) are prohibited or limited. Theocratic societies cannot withstand the challenge of belief systems where individuals are allowed freedom to reason and react in response to that reasoning. It is for that reason that the leaders of Theocratic type belief systems insist that their religious laws have dominion over civil governments so that laws can be enacted to provide the isolation needed to survive. To make the point, is it any wonder that there is so much strife and turmoil associated with countries where such opposing Theocratic philosophies are the law, especially when two such countries with opposing Theocratic philosophies have common borders or try to occupy the same territory as for example, Israel and Palestine? It bears reiteration. The problem is not that religious belief systems encompass conflicting philosophies with adherence chosen by followers through the exercise of freewill. The problem arises when conflicting religious philosophies are enforced upon people by prohibiting freedom to choose, through reason in response to freewill, which religious philosophies to follow.

It warrants repeating that only the belief system offered by Jesus exhorts love of God before all else. In that manner, benevolent beliefs are established in the conscious of individuals by freewill choice, which is necessary to influence freewill controlled behavior, which is the way our psyche is designed to work. As recorded in Matthew 22:37-40, Jesus said: "You must love the Lord your God with all your heart, all your soul, and all your mind. This is the first commandment. A second equally important commandment: Love your neighbor as yourself. All the other commandments and all the demands of the prophets are based upon these two commandments." Is it just by coincidence that Jesus said to love with heart, soul and mind since those are the mechanisms employed in the operation of the freewill behavior control system with which the Creator endowed humans for his glory?

The Christian Dilemma

There is another profound similarity between the Muslim and Judeo belief system and it has to do with their misunderstanding of the message of Jesus. Part of the blame rests with the followers of Jesus who even to this day wrap the explanation of details relative to the ministry of Jesus in myth, tradition, ritual, and parables, much of which are carryovers from the Jewish faith. Following the ministry of Jesus when the object at the outset was to modify the Judeo belief system, many of the Jewish traditions and beliefs were intertwined in the post Jesus church. Others were introduced in an effort to explain the ministry of Jesus and have been carried forward where they continue to muddle up the Christian belief system. It is hardly any wonder then that others, whose beliefs are aligned with the philosophy Jesus endeavored to change, are unwilling or unable to understand the message of Jesus and the details concerning his divine nature. To complicate the problem, Jesus used parables in an attempt to clarify his message for the many illiterate of the time but parables are easily misinterpreted even to this day. Jesus provided an explanation for the use of parables in Mathew Chapter 13.

Some of the misunderstandings involve beliefs that cause division as well as confusion, as was predicted by Jesus in Luke 12:51-52 where it is written: "Do you think that I have come to bring peace to the earth? No, I came to bring strife and division! From now on families will be split apart, three in favor of me and two against—or the other way around." Such a response is predictable since following the teachings of Jesus, people are being torn between opposing philosophical beliefs. If there was one thing that Jesus understood it was that even he could not force people to understand and believe for in Matthew 10:14 while instructing the disciples he said: "If a village doesn't welcome you or listen to you, shake off the dust of that place from your feet as you leave."

There are many trappings attached to the Christian faith. Some that may need to be retained as background history but are not necessary in promulgating the message of Jesus or of directing traffic on the straight way of Christianity. Some in fact provide grounds for misunderstanding and dissention. If the objective of a religious faith system is to bring people so that their behavior will be influenced by love of God, it stands to reason that disconcerting and discordant trappings that do not contribute to advancing that objective are a burden to the attainment of that objective. At least at some point, the essential beliefs should be presented free of trappings to the fullest extent possible. The following arguments are presented to illustrate the point:

- The Immaculate Conception and virgin birth. Jesus himself did not show interest in this matter. In fact his personal family was seemingly of little concern to his ministry since several of the Gospels (Matt.12: 48-50, Mark 3:33-35, Luke 8:21) quote Jesus as saying when told that his mother and brothers were outside and wanted to see him; "My mother and my brothers are all those who hear the message of God and obey it." Further, the Immaculate Conception and virgin birth can be explained using current knowledge of genetics without the need for invoking mystery and miracle. Since God was able to make new DNA for new life forms, surely he could insure that Mary's genetic structure carried the characteristics of King David that he wished to replicate in Jesus. Also, if man can in-vitro-fertilize an egg is it not reasonable to recognize that God could possess the same capability using the DNA of Adam for example? But the point to be made is that, while those explanations may have served a purpose in selling Jesus to the Jewish populous, perpetuation of such explanations serves no constructive purpose but now invokes misunderstanding, distraction, and discord.
- Jesus the Son of God. While potentially a controversial designation when taken literally, it is easily explained since God the creator of all things is in truth the Father of all and we, being children

of that creative effort, are in truth children (sons) of the Father. Anyone finding fault with that statement would do so solely to justify his objection to acknowledging the role of Jesus. However, that explanation falls short in explaining the special unique relationship that existed between Jesus and God. Jesus was and acknowledged that he was a real human being born of Mary. At the same time, a special spiritual interconnection with God the Father was established in the course of his baptism since following baptism and a successful test against the power of Satan, Jesus began his ministry. For anyone who acknowledges that God communicated with Abraham and with Moses through a burning bush, there should be no problem acknowledging God's ability to project his power and wisdom into Jesus who was especially conceived for the purpose.

- The Father, Son, and Holy Spirit are one in being. For those so inclined, it is easy to find objection to this claim but with a minimal explanation, those objections dissolve except for those who refuse to acknowledge that there is a God the Father the Almighty Creator of all. For those in that category, why worry about what they think. Jesus did not. Since God the Father exists in spirit form only and so is not limited by physical restrictions, God is able to project his presence in the form customarily designated as The Holy Spirit (or as an angel) which as previously explained was the mechanism employed in projecting his personality, power and wisdom to Jesus whose physical being was the son of Mary but whose spiritual being was an extension of God. Why is such an explanation objectionable if it demystifies the matter?
- Jesus the Messiah. This designation became controversial from the outset most probably in

deliberate rejection of the mission of Jesus. The Hebrew prophets had been prophesying for many generations the coming of a messiah in the form of a King who would free them of oppression and restore the glory of Israel as it existed under King David and who would rule forever. The Israelites could not and some still do not understand that the mission of the prophesied Messiah was to save the community of Israelites, then a Theocratic form of government that had repeatedly brought oppression upon the Israelites themselves, because the Israelite people were failing to obey God's laws. Again, the Israelites could not and some still do not understand that God sent Jesus, a descendent of King David as prophesied, with a new covenant that would influence their behavior and restore their righteousness and thereby the glory of Israel. It was the mission of Jesus to save the people of Israel and thereby the society of Israel from their oppression, which in their case was self-induced through their failure to obey God's laws. Either way, Jesus is their messiah whether they believe or not and now is, in the same manner, the savior of the world since the covenant he brought now applies to all who believe.

A Second Tower of Babel

"Come, let's go down and give them different languages. Then they won't be able to understand each other. In that way the Lord scattered them all over the earth—.—It was there that the Lord confused the people by giving them many languages, thus scattering them across the earth"[36]. According to this narration, the Lord deliberately confused the people at Babel so they would not be of one mind. When considering the fragmentation of Christianity following

---

[36] Genesis 11: 7-9

the defection of Martin Luther makes one wonder if that event may have been the Lord's way of getting people to look at Christianity from different perspectives as he did at Babel. If it was the work of the Lord with the intention of producing scattering, it certainly has achieved its goal but in the process caused some considerable confusion, division, misunderstanding, dissention and strife. It cannot be denied that the fragmentation that began with Martin Luther resulted in a thorough examination of all aspects of the then existing concepts and philosophies pertaining to Christianity. If the purpose of religion is to bring people into a faith/love relationship with God for the purpose of influencing behavior, has this examination resulted in improvement? All things considered, especially the growth in world population, there can be no doubt that the world would be worse off if religious practice remained as it was before the Martin Luther rebellion. However, considering the continued prevalence of immorality and evil, all indications are that there is need for continued effort. Demanding prayer in school or demanding the posting of the Ten Commandments will not do it. Beliefs cannot be forced and when attempted only cause dissention, divisiveness and ill will. Would a concerted effort to combat Relativism help reduce the prevalence of immorality and evil since Relativism permeates not only the Judeo/Christian/Muslim faith systems but also the Judicial and Civil Authorities around the world as well? You decide. If so, that effort must begin by acknowledgement of the truth concerning creation so as to strengthen trust in the belief in one God, the Father, the Almighty, and Creator of all things.

Religions Purpose

By dictionary definition[37], religion is an organized system of beliefs and rituals centering on a supernatural being or beings. That definition as it stands is a weasel worded relativistic statement since it does not define the purpose. The Catholic Encyclopedia goes to the extreme to provide an explanation of the term religion but also fails to

[37] Heritage Dictionary, Dell Publishing Co., © 1983

define the purpose. Is that just an innocent error or a case of using Relativism to achieve an ulterior purpose?

As previously explained, humans are created with a conscience wherein beliefs are stored. When deciding on a behavioral matter, beliefs that are the content of the conscience influence freewill. When the beliefs stored in our conscience reflect beliefs derived from the love of God our behavior will be influenced when there is trust and thereby faith in those beliefs. That process constitutes a faith relationship between the individual and God wherein freewill is not arbitrarily restricted but allowed to function freely. It should be recognized that the system for influencing the freewill decision making process is built-into the psyche by the creator so any attempt to create such a system anew or to attempt to alter that system, would be futile at best but most likely would interfere with the working of the God given system. From that it must be concluded that religion can have but one purpose, which is to attempt to draw people to an understanding where they will believe in and love The Creator God which will in turn provide influence over their behavior. From that point on, God will take care of the rest with a little help from the religious order to nourish and nurture that belief. The establishment of a community for that purpose with rituals for communal public worship of God is indicated to suit human nature but is not indicated as an absolute necessity. A problem is introduced when the rituals for worship take precedence and detract from the principal objective to draw people to believe in and love God. That is, to will the good of God's work which as stated, will influence the behavior of the believers.

All humans are equipped with the described system for influencing freewill decision-making but the system does not work until or unless individuals recognize and believe in the existence of an almighty Creator, the source of all power and wisdom. Even when activated through belief in God, the system is subject to malfunction especially when the meaning of the terms of the belief are liberalized through the application of Relativism so that the influence on behavior will produce behavior deemed by the individual to satisfy the conscious of the instigator. There are more instances of Relativism in religious practice than there is room to publish in one book but there are several where the intentions are very reveling. In one recent case, the body of a church condoned the ordination of a divorced and actively practicing homosexual pastor and then continued to condone his elevation to the position of a Bishop. Perhaps the most damaging of all, at least in recent years, deals with the movement to accept the argument that Jews, Christians, and Muslim should all just get along since all say they worship the same God even though there is a vast difference in the beliefs of each concerning God.

There is one belief involving both the Judeo and Islamic belief system that gives pause to wonder the purpose. Both belief systems hold strongly to the prohibition on the eating of pig while innumerable other restrictions on behavior of seemingly greater significance that were imposed through Moses, have been cast aside. The reason for such a restriction at the time that it was mandated is understandable since without proper handling pig meat was notorious as the carrier of deadly sickness. When such prohibition was lifted following the advent of Christianity, it makes one wonder why it was not only continued by Judeo practice but also then included in the restrictions on behavior mandated some 700 years later through Mohammed. Could that mandate have been retained to emphasize a distinction from Christianity so that the leaders of the Judeo/Islamic faiths could use that prohibition as an excuse to reject The Word of God delivered by Jesus? Or is it an indicator that Mohammed adapted the Moses laws in the beliefs he published. There is a second common belief that gives rise to the same question. Both allege to acknowledge Jesus as having stature at least equal to that of a prophet such as Moses or Mohammed yet neither acknowledge nor incorporate with any degree of priority the commandment mandated by Jesus to before all else, love the Lord your God. Could that have been a deliberate exclusion or just a case of ignorance of the implications? You decide but consider that there has been plenty of time to change if that exclusion was made in error. Since both belief systems acclaim the glory of the God of Abraham, by what reason if not with an ulterior motive, could the command of Jesus to love the God of Abraham be found objectionable? You decide since through the application of Relativism, other mandates of Moses and Mohammed were subsequently altered or even cast aside.

Chapter Ten: The Origins Argument

Framing The Argument.

For whatever reason, it has from the outset been the quest of mankind to understand creation and the creator. As time passed and understanding of the secrets hidden within the creator's work grew, a difference of philosophy began developing that resulted in the separation and specialization of theologian and scientific studies. It is that difference in philosophies between Theologians and Scientists that has become a stumbling stone to understanding the creator and the creation process. It is unfortunate but should be understandable that those beliefs that obfuscate and introduce obstacles to this quest of mankind to understand creation and the creator must be challenged even when those obfuscating beliefs have been introduced through religious beliefs.

In essence, Theologians have concentrated upon making their understanding concerning creation conform to the written outline of the creation process as documented in Scripture. They rely upon interpretation and reinterpretation of the written word of the Scripture and meditation thereon to reveal the nature of the Creator God. In other words, over these many years since the Genesis explanation was written, Theologians have made no significant contribution toward expanding understanding of the nature of the creator and creation or towards extracting the wisdom embodied in the Creator's work. Instead, the religious community has employed deceitful myths, mystery, ritual and tradition so that the wisdom hidden within the creator's work but not substantiated by Scripture does not need to be reconciled. Instead of acting to reveal the glory of God, they have aggressively acted to conceal God's glory.

At the same time, while the scientific community has made progressively advancing contributions to knowledge concerning creation and the creator through their continuing efforts to ferret out the wisdom hidden in the creator's work, they have closed their minds to the one factor that is key to their understanding. They refuse to acknowledge that creation was produced through the application of divine intelligence and wisdom along with energies, divine in nature, that have no physical component that produce forces that have no physical component. As a consequence, the scientific community has resorted to deceit by the propagation of mysterious theories as a means of explaining those processes whose understanding requires acknowledgement of that one factor they refuse to acknowledge. Instead of acting to establish the truth about creation, they have acted to conceal the first truth, which is in turn the key to understanding creation.

The unyielding positions of the theologians on the one hand and the community of scientists on the other hand has culminated in the development of a vast no-mans-land, a veritable chasm, wherein knowledge that is needed to understand the creator and creation has been pushed aside and replaced with doctrine and theories that hides the truth. The conflict of philosophies and doctrines pertaining to this critical subject is responsible for creating confusion, division and uncertainty in the minds of people and promotes the growth of Relativism since the truth is compromised to accommodate deceitful arguments, which are the roots of Relativism.

This then defines the opposing sides in this on going destructive argument. A certain segment of the religious community argues vehemently that any literal deviation from the summary of the creation story given in the book of Genesis in the Old Testament of the Bible can not be tolerated even though examination of that which was produced through that creation process provides proof that the Genesis summation, when accepted literally, is not technically correct. On the other hand, the scientific community and from them the academe community, already in a frame of mind to do so, adopted with open arms the philosophies espoused by Charles Darwin[38]. The Darwin theory was slanted to produce validation of the concept that there is no evidence of involvement of a supernatural all wise intelligent creator or super natural forces in the creation process.

[38] The Origin of Species by Means of Natural Selection, 1857

Establishing Alpha

There was a beginning and an understanding of the creator and the fundamentals involved in the creation process is possible only by beginning consideration from the moment in time, in fact the beginning in time, when the first physical matter was created. In the described on-going creation versus evolution argument, neither side has taken their consideration all of the way back to that moment in time when the first matter was created since to do so would prove both wrong. Therefore those arguments are flawed before they are made. The argument of the Creationists essentially begins approximately 10,000 years ago in the Garden of Eden when, according to their belief, creation began. Meanwhile, the Evolutionists take their considerations back only to the point in time that coincides with the creation of the first living organism estimated by the scientific community to have occurred approximately 6 billion years after the beginning of creation. The scientific community is satisfied to accept an undefined "big bang" as being the originator of the universe.

By starting at the beginning, it becomes evident that all physical matter in the universe is composed of atoms that are made of neutrons, protons and electrons, and that the assembly of the first atom came early in the creation process since the simplest atom consists of one proton and one electron. Even if there were an untold number of atoms created in a near unison "big bang", there was a first atom regardless of the process of manufacture and the intended characteristics of the end product as well as the components of which made were known before the first atom was made. ***That is a self-evident truth.***

It is known and acknowledged that there are forces involved in the assembly of the atoms of matter that cause them to function as they do, even though the Scientific Community has not precisely defined those forces except in theoretical terms. It is self-evident that these forces had to have existed prior to the assembly of the first atom and that a means had to be provided to cause these forces with the causing agent either in the form of an eternal energy source or provisions had to have been provided to replenish that energy as it is expended. It is also self-evident that the magnitude and characteristics of the energy involved is of divine nature since there was no other thing existing at that time and they have no physical component. It is further self-evident that the energy involved in creation could only be produced through the application of immense wisdom, foresight and intelligent reasoning because the precision, versatility and durability of those provisions made 12 billion years ago continue to serve the original as well as all new applications only now envisioned. Think about it, the energy that produced the forces to assemble and hold those first atoms is still in place but now expanded in the process of creation so as to extend to the ends of the universe and is the force that holds the universe together.

Scientific study with analysis of atoms of matter has established that the assembly of electrons, protons, and neutrons into atoms is extremely precise including the energy that provides the force used to hold their components together. Not only is the energy employed in assembly extremely precise with very unique behavioral characteristics that remain eternally constant at any point in the universe, each elemental particle was made to possess specific unique behavioral characteristics and each arrangement of elemental components produces very specific and unique behavioral characteristics. Every atom of a specific arrangement of components has precisely the same interchangeable behavioral characteristics as every other atom of the same arrangement but the characteristics of atoms change when the number of involved elemental components is changed. From what is known, it is self-evident that a precise and reliable methodology was employed to insure that specific behavioral characteristics were established and maintained consistent for each atomic element, which indicates the application of great wisdom and intelligent planning and that such sophistication was not and is not achieved by chance. Again, this is a self-evident truth and the energy that provides the force for establishing and maintaining the unique behavioral characteristics would out of necessity have to be extremely precise and eternal or provisions would be needed to replenish that energy as expended.

First Order Of Business

In the beginning, either the creating intelligence knew in advance or, after assembling a few atoms of matter, soon realized that an "affinity" force was necessary so that the elemental components that were to be assembled into atoms and the atoms themselves would not fly off and become lost forever. Such a requirement would have been necessary regardless of the method of creation but especially if an explosive force such as the big bang were employed. However, since such an affinity force is a prerequisite for assembly of the elemental components of that first atom, it is self-evident that such a force existed prior to beginning assembly and that force is very precise with unique and stable behavioral characteristics. Regardless of the mechanics of manufacture and assembly, it is self-evident that great wisdom and intelligence was involved to recognize and devise a means for establishing an affinity between the various components involved with creation as well as with the "source" to which all matter is anchored. Although having no physical properties and incapable of measurement except by observing its behavior on physical bodies, the community of scientists does in fact recognize the existence of such force even though having no physical properties, and has chosen to call that affinity force "gravitational attraction". Unfortunately, rather than acknowledge that there is an energy that is the cause of gravitational attraction, the community of scientists theorize that gravitational attraction is produced by an "affinity" that exists naturally in each elemental particle of matter. The term "affinity" is left to stand undefined. Through that explanation the scientific community is eliminating the need to recognize and account for a causing energy, which would require confronting the existence of a divine cause. *Such an admission would be contrary to their avowed philosophy that denies the existence of an all-wise intelligent creator.*

Denial by the scientific community of the existence of an energy source responsible for producing gravitational attraction is the cause of their inability to understand the processes involved in the creation of the universe. As previously stated, the force the scientific community calls an "affinity" would of necessity have to be eternal or provisions would have to be in place to maintain that force to extremely close tolerances. Even a minute deviation in the strength of that force would cause the universe to implode or explode. Also of necessity would be provisions for adjustments in strength of the affinity force to accommodate changes in the amount and distribution of matter throughout the universe, especially to compensate for mass lost through conversion to energy.

Fundamental Force Characteristics

The fundamental force called gravitational attraction, one of four fundamental forces involved in the assembly of the universe according to Albert Einstein, has been recognized, investigated, and analyzed over thousands of years even to the extent of developing a formula that theoretically will calculate the power of those forces. Because gravitational attraction acts upon physical objects where the effects can be observed and measured, considerable has been learned. For example:

- It is known that the strength of the gravitational force is extremely precise and that the strength of that force as measured at the center of gravity in a particular assembly, is in proportion to the amount of material involved; that is, one unit of strength for each unit of mass.
- It is known, or should be recognized that protons are the smallest particle capable of having or of serving as a center of gravity. Neither electrons nor neutrons have "affinity" power since they do not independently attract other particles. Therefore, it stands that protons provide the mechanism through which gravitational attraction operates. Each proton is supplied a quantity of gravitational attraction energy to match the mass of the proton plus the mass of all satellited electrons and neutrons as well as the energy involved in establishing and maintaining the connection between the center of gravity and satellited matter.

- It is known that in an assembly of a number of proton/neutron/electron assemblies, the center of gravity of each such assembly is supplied a quantity of gravitational attraction with strength equal to the sum of the strength of all assemblies satellited to that center of gravity, as well as the energy involved in establishing and maintaining the connection with satellited matter, which strength changes as the number of elemental particles in the assembly changes. Further, it is known that the strength of that gravitational force diminishes with distance from that center point of the particle or assembly under consideration.
- It is known that gravitational force acts radially from the center or core of each individual proton as well as from the center or core of an assembly of elemental particles where in every instance the direction of force is always attracting or pulling toward the center or core of the assembly under consideration.
- It is known that gravitational force acts at the speed of light (186,000 miles per second) through any medium, it cannot be deflected at least by physical matter.
- It is known that the force called gravitational attraction, also known as the force that produces gravity, has no physical component. In other words, the gravitational attraction force that pulls upon a physical body and causes that physical body to be moved does so without establishing an apparent physical contact or an apparent physical connection. For example, the gravitational attraction acting out of the core of the sun acts to

> hold the earth and other planets in orbit without establishing a physical connection.[39]

Problems With Current Creation Theory

While the preceding behavioral characteristics are known, the scientific community has closed their mind to the idea of acknowledging the existence of a source of energy responsible for producing gravitational force. It is not sufficient to pass-off the cause of the force known as gravitational attraction as just a natural characteristic of matter. Even if gravitational attraction is an inherent characteristic of each particle of matter, there must be energy to produce such force and there must be a source of that energy. One of the basic laws of physics recognizes that the application of a force requires the expenditure of energy. It then follows that there must be a source to supply the energy as it is depleted or if internal, the source would have to be eternal with no diminution over time. Considering the miniscule size of particles such as protons that are invisible even with a powerful microscope, and the length of time since creation now estimated at 12 to 13 billion years, it is a tremendous stretch on the imagination to accept that the energy to produce such precise force could be stored internally, especially in the absence of any proof.

Since it has been verified and acknowledged that the force of gravitational attraction of a body or an assembly of particles acts radially to attract by pulling particles or objects within range toward the center or core of the body, it stands that the energy source for the gravitational force of a body is located at or acts from that core point of any assembly including the assembly known as the universe. That conclusion must be acknowledged as **a self-evident fact**.

Also, since the current scientific theory does not acknowledge an external source for the energy that produces gravitational attraction but attributes the cause of gravitational attraction to "an affinity" inherent in each elemental particle of matter, that theory is saying that

[39] The method by which accomplished is discussed in the book by the author, All Things Seen and Unseen ©2002, 1st Books.

the source of the energy producing gravitational attraction force resides within each elemental particle of matter. It is then also saying that the gravitational attraction force radiated from the center of an assembly of particles, say the earth, is produced because the "inherent affinity" of the involved particles is combined to act from the center of the assembly with a strength equal to the sum of all particles in the assembly. **Immediately three ambiguities are revealed**. One: how can each particle retain its inherent "affinity" yet transfer that "affinity" to the center of that assembly and then to succeeding assemblies, even if a mechanism existed to facilitate such transfer. Consider that the "affinity" of a proton in a body standing on the surface of the earth is transferred first to the body, then to the center of the earth, then to the sun, then to the center of the Milky Way and so on. Two: if, as has been established, the strength of gravitational attraction diminishes inversely as the square of the distance over which it is acting, how can the strength of the gravitational attraction of particles at a distance from the center be transferred over that distance and caused to act from that center point without diminution of strength? Because of this fact, the transfer of gravitational attraction to the center of an assembly would not be in direct proportion to the number of elemental particles of which the assembly is made and therefore is not in agreement with acknowledged fact. Given that scenario, there could not be one unit of "affinity" force at the center point of a body for each unit of elemental particles as is stipulated by current theory. The preceding concept of affinity force transfer becomes even more ridiculous when considered on the scale of distances that exist between planets and stars and between stars and galaxies. Three: Contrary to the current theory, electrons and neutrons do not possess the ability to attract other matter yet it is known that the center of gravity of an assembly that includes electrons and neutrons possesses gravitational attraction strength equal to the mass of the assembly including all electrons and neutrons. From this it is clear there must be an external source for the energy that produces the gravitational attraction that acts through the center or core of a body.

Flawed Science

While the previous argument is sufficiently convincing as to establish the fallacy of the current theories expounded to explain gravitational attraction, there remains one as yet unrecognized fatal flaw in the present theory. The infamous Einstein $E=Mc^2$ equation must be defective since that equation considers only the energy of which matter is made. It does not take into consideration the energy involved in creating and maintaining the "centers of gravity" of a chain of assemblies of matter nor the energy within the bonds of gravitational attraction interconnecting those centers all of which are in fact formed of the same energy of which matter is made. The energy embodied in those centers of gravity and the bonds interconnecting those centers cannot be ignored. Consider a simple case where an electron is ejected from its orbit around the nucleus of the atom. The energy involved in establishing and maintaining the bond with a proton in the nucleus is thereby fragmented and released as heat or light. If, as is the case with a 100-watt light bulb, the approximately 6,240,000,000,000,000,000 such bonds that are broken each second are responsible for a considerable portion of the energy released. Consider then the magnitude of energy involved with the bond between a proton and a neutron where the mass of matter is in the order of thousands of times greater than the electron bond. Could that unrecognized energy explain "black holes" and the "missing mass"?

There is a long-standing law in Physics called the law of the conservation of energy that deserves to be revisited. Stated as the most important of all of the laws of nature, it proclaims in one form that *energy can neither be created nor destroyed.* It is sometimes stated that *when transforming energy from one form to another, energy is always conserved* and also; *the sum total of all energy in the universe remains constant.* But something doesn't add up since it has been established that the universe has been expanding both in size and amount of matter over the estimated 12 billion years of its existence. **From this it must be concluded, there has existed throughout this period of time, a source for the newly created energy and matter that would be needed to sustain this expansion.** Other than to try to deny creation and a creator, is there a valid reason why this information should be suppressed?

Defining Heat

When it is recognized that gravitational attraction is the manifestation of the same energy of which physical matter (electrons, protons and neutrons) is made, an accounting for that energy must be made. When an accounting is made for that energy, the mystery surrounding light, heat, and electricity, chemical, atomic and nuclear energies becomes explainable. When electrons, protons and neutrons are fragmented, the energy of which made is seen as heat when released, as is the gravitational attraction energy that produces the forces involved in their assembly into atoms. For example, when released, the energy that holds electrons in orbit around the nucleus of an atom is seen as heat when the electron orbit is disrupted such as occurs in chemical reactions or in the flow of electricity in a conductor. Greater quantities of heat are released when electrons, protons or neutrons are themselves destroyed such as occurs in combustion, and atomic/nuclear power release.

Properly defined, heat is the name given to the energy remnants released when any created matter, seen and unseen, is destroyed. The character of heat is to expand until it achieves neutrality, which corresponds to absolute zero degrees or nothing. Released heat is destructive often causing the failure of other orbiting electrons that in turn release more heat and under the proper conditions may produce combustion or even explosion. When exposed to heat, cellular tissue is destroyed whereupon, being so programmed, our brain registers the sensation of heat. It was the destructive nature of heat that precipitated the use of cold as a means of slowing the decay of perishables and experiments with cryogenics as a means of modifying the behavior of atoms. Heat has been used as a means of destroying bacteria since prehistoric times even though the process and reason were not then thoroughly understood. So it is today. Our understanding of the process involved in creation has been flawed by the failure of the scientific community to recognize and acknowledge that gravitational attraction involves the same energy as was employed in the creation of matter. Unfortunately it provides evidence that the books now in use by the Academe Community of the world are flawed, as is the education of those who have been educated using that flawed material.

Alternative Explanation

There exists a very simple explanation for the mechanics of gravitational attraction. When it is acknowledged that the force of gravitational attraction has no material or physical component, has no weight, occupies no space and is not interfered with by anything except the passage of time and thereby distance, it is self-evident that gravitational attraction is of divine nature. Also, when the first principle of physics is acknowledged that says a source of energy is necessary to produce force, which is work, it is self-evident that a source of energy is necessary to produce gravitational attraction force. Further, since it is already acknowledged; that the direction of the pulling force progresses from the smallest particle through succeeding larger bodies and in every instance toward the center of each assembly in order of size towards the center of the universe; that those forces are divine in nature including the ability to travel through space without the need of a medium, but which in fact establishes a medium; that of necessity a source of energy exists; that the intensity of the force of gravity behaves in conformance with the inverse square rule; that the strength of the gravitational attraction force at the center or core of a body is in proportion to the number of elemental particles which equates to the mass of that body; then ***it is self-evident*** that the source of the divine in nature energy that powers gravitational attraction exists at and is radiated from the vector center of the physical universe. ***It cannot be denied.***

There is another way to visualize the mechanics of a core-centered universe. It begins with recognition that there was a first sub-elemental particle of which particles called Quarks are made and then subsequently Quarks are assembled to make neutrons that exhibit physical characteristics. Since such sub-elemental particle are made out of a form of energy as acknowledged by Albert Einstein, that particle was not only connected to that source of energy in the course of manufacture but in fact remains connected not only as a means of distributing "affinity" energy to each particle but through which connection "affinity" force (gravitational attraction) is applied. Succeeding particles produced at and likewise attached to that central energy source point through a gravitational attraction connection, with each particle supplied one unit of affinity power, surrounded that center point in a radial manner. As sub-elemental particles were assembled their gravitational attraction connection were routed through the "center of gravity" of each such assembly, which point acquired an "affinity force" equal to the sum of the units of sub-elements involved in that assembly. As more and more particles and assemblies of particles were produced, expansionary forces, whether centrifugal forces, like charge repulsion or the heat of an explosion, caused the particles to be moved further and further from the point of origin, which is to say that the universe grew out of its center with the gravitational attraction connections "stretching" as the universe expanded with the strength of attraction force diminishing as the distance from the source increased.

So it was when particles assembled into bodies. Each assembly began with one particle, which became the center of the assembly. As other particles were attracted to the assembly by interacting gravitational attraction forces, each became satellites of the originating center point of that assembly whereupon, the strength of the gravitational attraction at that point became one unit of strength for each elemental particle attached to the assembly. The center point or center of gravity of each independent assembly becomes both the anchor point for the force of gravitational attraction, as well as the energy distribution supply point, for the individual satellited particles or assemblies of particles. That is why the strength of gravitational attraction force at the center or core of an assembly such as the earth or sun is equal to the number of individual elemental particles satellited to that point of the assembly, which equates to the mass of the assembly. As previously explained, the strength of the attractive force in an assembly diminishes as the distance from the central core point increases. That is why the gravitational force at the core of an assembly, such as the earth for example, is stronger than at the periphery. (An analogy to explain the loss of strength with distance phenomenon is presented in Chapter 12.)

Without clear contradictory evidence, it makes better sense to base physics and the circumstances pertaining to creation of the universe on ***self-evident truths*** that have in fact been established based upon scientific study, than upon agenda driven, misleading, unsubstantiated theories and traditions. In so doing, it is not necessary to acknowledge that the source of energy so critical to the existence of the universe has any connotation other than reality. Acknowledgement leads to an understanding of those factors concerning creation that mystified Albert Einstein and most other Physicists, as well as Theologians.

Chapter Eleven: The Evolution Argument

Framing the Argument.

The argument set out in the previous chapter frames this argument as well. The principal point of this argument arises out of the theory advanced by Charles Darwin to the effect that species of life originated and evolved through natural selection[40]. The Darwin theory was slanted in an attempt to validate the concept that life originated and species evolved without the involvement of a supernatural intelligent divine creator or super natural forces. The previous chapter provided substantiated evidence that a force called gravitational attraction with no physical component that is powered by a form of energy with no physical component is not only the key factor in the creation of all physical matter in the universe but provides the mechanism by which the universe is powered. The universe could not exist without the energy source that produces the "affinity forces" called gravitational attraction. Any attempt to explain the origin of the universe and life is invalid and without merit except or unless that matter is first addressed and reconciled. Such is the case with the theory of the origin and evolution of species through natural selection as is currently being taught exclusively in schools throughout the world.

The Proton/Electron Relationship

While the energy that produces gravitational attraction is the key energy involved in the creation of the physical universe, there is a variation of that force that is the key factor in the assembly of elemental particles into the more and more complex organic and inorganic assemblies that give form and utility to the universe. The assembly of atoms and all subsequent assembly of matter are accomplished through the use of and are dependent upon this

---

[40] The Origins of Species by Means of Natural Selection, Charles Darwin, 1857

variation of gravitational attraction. Nothing except elemental particles would exist without it. The creation and evolution processes cannot be understood until and unless the nature and behavioral characteristics of this force are first acknowledged and understood.

Particle Physicists have documented that *neutrons*, when not united with a proton as in the nucleus of an atom, are not stable and with the conditions existing on earth, *degrade* in approximately 20 minutes *to produce a proton and an electron.* ***That fact provides evidence that creation began with the assembly of neutrons that "degrade" to produce protons and electrons and then proceeds to the assembly of atoms.*** The process of producing protons and electrons involves the constriction of neutrons, in response to immense core centered gravity driven forces, (especially as would exist near the center of the universe) to a point where one of the elemental sub-particles of which neutrons are made is ejected from the neutron in the form commonly called an electron. Following the ejection of an electron, the remnant material, vastly constricted, forms what is commonly called a proton. Physically a neutron is many times larger than a proton even though having only one additional electron sized particle, which provides an indication of the degree to which protons are constricted. Having been constricted sufficient to eject an electron, the proton is left with a one-electron deficit along with the energy source and power that initially served that electron. Other than that, neutrons and protons are physically identical in composition but vastly different in behavioral characteristics. With that residual unit of energy, protons emit a force endeavoring to acquire an electron to fill that deficit that exists in each proton. However, the core-centered forces that produced the constriction that caused the electron to be ejected, now holds the proton in its highly constricted form, which prevents re-absorption of the electron back into the assembly which action, if it could be accomplished, would result in the recreation of a neutron. (Physicists would recognize that energy was expended in the process of constricting neutrons to produce protons which energy would need to be replaced if the proton where to be re-expanded and Physicists would recognize that energy to be what is now called "nuclear energy".)

The residual proton/electron "affinity" produced following the ejection of an electron from a neutron constitutes a unique variation of gravitational force that in physics is called ***electrical energy****. That proton/electron affinity*[41] *exists between a proton and an electron only* and in that capacity produces the force involved in chemical reactions (including oxidation), in electricity, in electro-magnetics, in electric motors and generators, and in all bio-processes including metabolism and contraction of muscles. The sequence of events just described is worthy of special note by Physicists, Biologists and Theologians since it is the earliest evidence that "built-in logic" is involved in and plays an important role in creation. *This event is further worthy of special note since it illustrates that* ***the action described is driven by the behavioral characteristics built into the energy of which all things are made***, known generally as the energy that produces gravitational attraction.

Inorganic Evolution.

All things physical in the universe are assembled in building block fashion from elemental particles. As previously stated, neutrons, protons, and electrons held together with the force of the proton/electron relationship called electricity, are assembled into atoms of the various elements of matter. Each electron, each proton and each neutron have unique precise behavioral characteristics, no matter where in the universe they are used, that is set by their interaction with gravitational attraction and the proton/electron relationship. Atoms take on a unique precise behavioral characteristic depending upon the number of protons, electrons, and neutrons in the assembly so that a number of the approximately 75 (that exist naturally) different active types of atoms, with each type possessing unique precise behavioral characteristics, serve as the building blocks of all matter in the universe. The assembly of electrons, protons, and neutrons to produce atoms is beyond human ability and the discussion of that process is beyond the purpose of this writing. Atoms were

[41] This subject is presented in greater detail in the book by the author "All Things. Seen and Unseen".

joined or are made to become joined, to produce molecules and compounds where each combination acquires a precise unique behavioral characteristic, which then serve as building blocks for more complex combinations. Since each arrangement of components produces a unique behavior, it follows that the unique behavior characteristics can be employed to create built-in logic capable of initiating action not unlike microprocessors. Everything physical in the universe is produced through this building block process, which also creates innumerable instances where built-in logic exists. According to current estimates, that process progressed for about 6 billion years before an environment was created capable of supporting the development of living organisms especially on the planet earth. ***The process called evolution*** when intelligence directed as applies to creation, ***cannot be denied.*** Those who attempt to deny evolution processes render their argument mute.

Beginning Of Living Organisms

As with inorganic matter, there was a first living organism and as is the case with the development of physical inorganic matter, living organisms are an assembly of elemental building blocks. The basic building block of living organisms is called a biological cell but unlike the atom made of simple elemental particles, living cells are made of complex compounds and molecules that in some cases exhibit exotic and complex built-in behavior including execution of logic and the production of physical activity. These cells are living organisms. Many of the functions of even the most basic living cell depend upon the behavior of one complex assembly controlled by built-in logic acting upon and directing the behavior of other complex assemblies that themselves respond under the direction of their own built-in logic. Each of the various sub-assemblies vital to making a living cell are an assembly of specific atoms and molecules aligned in a specific precise manner so as to produce a specific vital behavior that is extremely precise, reliable and unique to that specific arrangement. *These vital sub-assemblies are not just a jumbled assembly of atoms and molecules* and they are also not by themselves living organisms. The selection of components used in each case and the arrangement of their assembly in each case is the determining factor in how these assemblies will function and what function or functions they will *perform* ***when assembled in a cell***. In other words, *the first living cell could not have been made until and unless the necessary complex sub-assemblies, that are* ***not capable of living and replicating independently outside of the environment of a living cell****, were available in working order.* **Therefore**, in order to assemble the first cell capable of living and perpetuating life in the environment then existing, **all of those vital complex sub-assemblies had to have been especially created and in working order at the precise time and at the precise place that the first cell was being assembled.** Remember, when the first living cell was made there were no cells existing to produce those vital complex sub-assemblies and they do not and did not exist in nature.

Consider the complexity of these sub-assemblies vital to the making of a living cell. First there is the skin of the cell and of the sub-assemblies; then the mechanism for extracting nutrients and converting the nutrients into energy; then the DNA mechanism encoded with the instruction on how the cell is to function including assembly, gathering nutrients, making energy, and reproduction including how to replicate these complex vital sub-assemblies; then the mechanism that reads the DNA and converts those instructions into actions. The interdependent complexity of the processes involved in making that first living cell capable of living and then reproducing to perpetuate the chain of life, involves so many interdependent vital processes that such an operation solely through chance is absolutely ridiculous especially since the contents of living cells give no indication of failed efforts. And, it is of no use to assert that perhaps the first life on earth came from some other world since there still had to be a first living cell with the listed vital complex sub-assemblies manufactured and assembled regardless of where that happened.

In concluding this discussion, it is important to recognize that the assembly and powering of atoms and molecules and the establishment of the behavioral characteristics of those assemblies and sub-assemblies, ***is in all instances made possible because of the proton/electron relationship that was produced with the conversion***

of the unstable neutron into a proton and an electron. *In other words, the energy that powers the functioning of cells as well as the physical assembly of the cells is provided through the proton/electron relationship that is a sub-system of gravitational attraction.*

Evolution of Living Organisms

The practice of individuals and certain religious groups taking offense over use of the term evolution is foolish and counter productive. Evolution, a gradual process in which something changes, especially into a more complex form through progressive development, is an inherent characteristic of creation. Evolution was a factor in creating the physical universe that began with that first atom and progressed over the estimated 6 billion years until an assembly of matter was developed with the combination of factors that would permit the development and sustenance of an environment capable of supporting and sustaining living organisms and then for another estimated 6 billion years for the progressive development of living organisms. Even the story of creation in Genesis includes the progression of development over 6 "days", which in fact constitutes evolution.

The progressive development of the physical world would have been somewhat in the order of mundane backbreaking labor since it involved the manufacture of a massive quantity of atoms of matter, their assembly into the physical bodies and the distribution of those bodies throughout the universe in the form of galaxies, suns and solar systems with orbiting planets. All evidence indicates that the atoms of matter that had been spread throughout the universe, collected into assemblies of galaxies and solar systems with planets in response to the forces of gravitational attraction. With assembly under the influence of gravitational attraction the tendency would be for particles, at least initially, to stratify with denser particles closer to the core of the assembly whether in a gaseous, liquid or solid form except where particles of like charge are held in colloidal suspension, which involves forces produced by the proton/electron relationship. It is theorized that an assembly of protons (or hydrogen) at absolute zero temperature, would produce the densest possible body and thereby the most concentrated gravitational attraction force.

The progressive evolutionary development of living organisms gives the appearance of being a considerably more technically dramatic, interesting and exciting process than the creation of the physical universe, especially when it is recognized that living organisms, behaving as willed by the creator, were the instruments employed by the Creator to make those changes in the environment. To begin with, the evolution of species had to have gone hand in hand in a deliberately coordinated manner with the development progressively of an environment needed to support the progression of species. *Contrary to theories advanced by the Darwin Evolutionists, living organisms did not change to fit the environment.* The process of providing a suitable environment was deliberately progressive since *it was the living organisms at each step that were the instruments used by the Creator to produced progression in the development of a better environment needed to support a more sophisticated life species.* ***In living, each species of life produced changes to the environment. Through dying, they produced further changes to the environment.*** *Knowledge of what organisms would live in the then existing environment and then what organisms were required to produce desired progressive changes in the environment, would have been required before-hand in order to avert the creation of an environment that would have killed then existing life and terminated the perpetuation of the life process.* Thus it is seen that there was a purpose for the various species of life. The Darwin theory of evolution through natural selection is bogus, without basis.

Although the process of environment and living organism evolution is complex and time consuming, once the complicated, complex, interdependent, vital cellular sub-assemblies were perfected and assembled into the first living organism, the mechanism involved in the progressive development of species is a relatively simple operation controlled by the genetic code written in the DNA. The difficult aspect involves determining and coordinating the improvements in the environment required to support the next step in living organism development and then the "writing" of the new code needed to produce the new features needed to effect evolution of life forms and the environment. It is interesting to realize that some code prepared for the first living organism may still be in use in the DNA of current day life. *This is the difficult, complicated, complex aspect of this process and essentially impossible without the foresight and wisdom to know what is needed (desired) and how to write the new DNA code necessary to modify life forms to achieve those objectives.* This is the process involved in evolution. ***It is important to recognize that with each progressive step, the objective is the attainment of a specific predetermined behavior.***

It is not difficult then to realize that any mistake in the coordinated progressive development of the environment and living species would have destroyed *the one component vital to the evolution of both, the living cell needed to produce a new life.* Considering the number of steps in this progressive process since creation of the first living cell, and realizing that non-lethal mistakes and defects in DNA coding would also be perpetuated along with other code modifications, the degree of planning and coordination could not have been produced by chance or by natural selection. It is important to realize that the coded instructions that tell cells how to behave and what actions to perform especially in the area of reproduction, include provisions that requires the melding of the genetic characteristics of the parents in their offspring but inhibits the melding of genetic characteristics that cross species lines. Therefore some melding in the physical and personality features within the limits provided for in the DNA code, occurs in the process of reproduction but such melding does not involve the writing of new code for new features as would be required for the creation of a new species. If those melding provisions constitute "natural selection" then so be it since that is the way the writer of the DNA code intended. **There is one thing for sure**. *A brainless single cell organism was and still is incapable of the intellect, wisdom and physical agility needed to recognize what new features and environmental changes were or are needed, let alone the ability to write and introduce the new code, to achieve such changes.* Even with today's level of education, knowledge and computer sophistication, such a fete is not possible.

Interdependent Complex Vital Mechanisms

Definition of those complex system that were vital to the development of that first cell and every cell since, is beyond the scope and purpose of this writing. For those with an interest, information on those systems can be obtained through a more in-depth study of the subject[42]. It is the purpose of this writing to show the processes involved in creation and to point out some of the inconsistencies and impossibilities embodied within the theories and claims of the opposing sides of the evolution/creation argument.

It deserves repeating. There are complex systems within cells that are vital to making a cell a living organism capable not only of sustaining that status but also capable of reproducing those vital complex systems, which systems are available only through the reproduction of "the mother cell". They are not available pre-assembled in nature. But there was a first cell created at a time when there was no "mother cell" from which to inherit those complex systems that are incapable of sustaining their viability except within a functioning living cell. There is only one answer for how these complex inter-dependent systems came into being all at the same time and all at the same place on this world and all ready to function with each other in a delicately balanced organism, suitable for the then existing environment, that were then properly assembled to became the "Eve cell" from which all subsequent life grew.

There is a significant self-evident fact involved in the assembly of that first cell and all subsequent cells. Those complex vital sub-systems cannot function with the degree of coordination and purpose necessary to form a viable functioning cell except and unless operating in response to specific, relevant, directing instructions, that are provided by being encoded on the DNA. DNA is an extremely sophisticated, durable, micro-miniature, complex assembly of extremely sophisticated molecules each with extremely precise behavioral characteristics. Those molecules are arranged in such a way as to produce "written" instructions not only to instruct each cell

[42] Darwin's Black Box: The Biochemical Challenge to Evolution, Michael Behe, The Free Press, 1996.

how to function to obtain suitable nutrients but how to convert those nutrients into energy and what to do with that energy. The DNA instructs how and when the various sub-assemblies should function, how and what to communicate with other sub-assemblies, how and what to communicate to other cells and what behavioral characteristics the cell, and even an assembly of cells, should possess. At this point, it makes no difference how the various cellular sub-assemblies came into existence initially, although previous consideration in this manner supports the argument that they were deliberately designed for their purpose. It is however, ***conclusively evident*** that the DNA could not have been assembled in a manner to provide such intelligent instruction except or unless through the application of extreme intelligence and unlimited wisdom. Among other things in addition to knowing how to write DNA code, the writer of the code had to have a complete understanding of how and what it takes to make each of the sub-assemblies as required in making each specific type cell. Cells are not just an assembly of complex sub-assemblies and components but they must function in a coordinated manner to achieve specific results. Consider the intercellular and intra-cell coordination involved in causing a person to walk or talk or see which complexity had to have been envisioned before the first cell with DNA was created.

Summary of the Evolution Process.

The evolution of life process began with that first cell that was equipped with a set of instructions capable of telling that cell how it is to be assembled, how it is to function, and how it is to reproduce itself including the components of which made. From that point forward, the progression in development is achieved by inserting new code designed to produce the new features or behavioral characteristics as well as the desired environmental impact. It is relatively easy to recognize at this point in the evolution process that new code must have been inserted into the DNA of that first cell when a new species with new features or behavioral characteristics were to be produced since the needed code would not have been available by any other means. Again, the described operation means the application of extreme intelligence and unlimited wisdom.

It is important to remember that the forces involved in the assembly and functioning of cells are produced by the proton/electron relationship that is a variation of the force seen as gravitational attraction that is produced by the energy of which all things, seen and unseen, are made. It is also important to remember that the built-in behavioral characteristics of the components from individual atoms to complex molecules and assemblies of molecules, was produced out of the characteristics of the force seen as gravitational attraction that is produced by the energy of which all things, seen and unseen are made.

Defining Life.

The definition of life, that which causes an organism to be alive when present but dead when not present, has been shrouded in mystery and myth. Theologians have taken a position that life is a "spiritual in nature" energy that is bestowed upon a gamete cell by the Lord, the giver of life, at the time of fertilization. Some have speculated that the energy that gives life, as in the case of mammals, may be transferred in the sperm donated by the male parent. The mystery concerning the nature and source of the energy that gives life has deepened following demonstration in recent years that the principal agents involved in reproduction consisting of sperm, egg cells, fertilized eggs, and even embryos, can be frozen for storage and then reactivated to complete the reproduction process. The process of cloning has added to the confusion even though it is that process that has helped to open the door of understanding exposing the nature of life.

Actually, the definition of life was revealed in the previous discussion. Life began with the creation of that first living cell that had everything necessary to sustain its living status plus the ability to reproduce itself, including all of the internal assemblies vital to the life process, to produce a new self-sustaining living cell. *Since that first living cell was created, all that is involved in producing a new life is the reproduction of an existing living cell* ***by the processes***

provided for in the existing cell. From that point, even progressive evolution development is achieved by the insertion of new DNA code into a cell that is otherwise complete and capable of living and reproducing. All life proceeded from that first cell that has been multiplied through reproduction to produce all intervening as well as all existing life.

Consider the following arguments. (1) In normal conception using mammals for example, all of the components needed to produce a new living offspring are included in the egg cell produced by the mother, *except that the DNA code needed to direct the functioning and development of that cell is not complete. Once the DNA code contribution of the father has been "married" with the DNA code contribution of the mother that is already in the egg cell,* **that egg cell begins living as a new offspring**. *Remember that the energy and forces needed for a cell to live, to be alive, are provided through the proton/electron relationship and the mechanisms necessary for that process to work are included in the egg cell produced by the mother.* ***All that is needed for that egg cell to begin functioning is the completion of the DNA code.*** (2) With freezing, whether egg, sperm or embryo, the proton/electron functions are slowed down by the removal of heat. Heat is a factor in the fragmentation (oxidation) of electrons which is a process involved in the proton/electron activity. (3) With cloning, the nucleus of ***an already living cell***, *where the marriage of the maternal and paternal DNA contributions has already been completed*, is inserted into an egg cell that needs only a completed DNA code to function. In fact, the reproduction of cells in the normal progression of growth and life maintenance is a cloning type operation.

In the vernacular of a recent political campaign slogan: "***it's the cell stupid***". It bears repeating. **The mystery of life rests with the creation of that first cell made complete by the Creator with all of the mechanisms necessary to live, including those mechanisms necessary to replicate itself to produce a new functioning living cell that is itself complete with the ability to replicate to produce a new functioning living cell.** That is, self-perpetuation. In the case of multiple celled organisms such as a human, growth from a single cell is accomplished through the reproduction of cells as previously explained while the process of producing a new offspring is accomplished using a special "egg" type cell. Egg type cells produced by the special organs with which females of the species are equipped, are complete and ready to begin functioning except for its DNA code. The DNA code of the egg cell is made complete and ready to begin directing the otherwise complete cell's function with the marriage of the paternal DNA contribution to the maternal DNA contribution that was contained within the egg cell, as previously explained. When the DNA of the otherwise complete egg cell is not completed, the egg cell cannot begin functioning since it takes a completed DNA to direct the activities that are recognized as making the cell alive. After the DNA is made complete the cell then is complete and thus able to function to complete the development of a new offspring. By this explanation it becomes evident that ***the egg cell***, powered by its many proton/electron relationships, and in the case of mammals made complete by the marriage of the maternal and the paternal DNA code contributions, ***is the carrier of life to new offspring***. The life that was created by the assembly of that first cell and which has been nurtured through all of the many intervening generations is extended in each case through cell reproduction to give life to new offspring. For the love of God! Such is the glory of God.

Sanctity of Human Life

Given the preceding disclosure, it would be a dereliction of integrity if the sanctity of human life factor where not discussed. Under any circumstances, there can be no doubt that the creation, perpetuation and advancement of life is and has been a, if not the, principal objective of the Creator. Beyond what is now discernible and understood through the study of that which the Creator has made, it cannot be known with certainty what the ultimate objective of the Creator is however, the establishment and maintenance of control over behavior so as to assure the perpetuation of creation has been a principal consideration throughout the creation process. With regard to the subject, certain things are obvious. For one, life apparently was not initially considered absolutely sacred since living things were, even prior to Adam and Eve, designed and therefore assumed as intended to have limited life spans because those provisions are built-in. Also, it is apparent that components vital to reproduction such as the carrier of life, the egg cell, and sperm, the key to heredity, where treated as expendable since provisions were not made for the conservation of those components. That is, there are far more egg cells and sperm produced and sacrificed than are fertilized to perpetuate life. That said scripture records that the Lord did not take lightly the killing of Abel by Cain. Also, following the destruction of life by the great flood initiated at his own hand, his covenant with Noah included the command that murder is forbidden whether by man or animal.

It has been the general consensus of humanity for the period of recorded history, especially from a theological standpoint, that human life is sacred although instances of human sacrifice as well as the sacrifice of human life in military conflict have been commonplace even under the guise of religious beliefs. There is to say the least, lack of consensus as to what is the life giving part of a living thing. Certainly, many of the long held beliefs on the matter have been overtaken and rendered fallacious by subsequent growth in understanding through biology including recognition that there are in fact many things that are vital to the sustenance of life. If, as proposed in the preceding discussions, it is the egg or seed that is the carrier of life from which a new living thing is produced. Does that make the egg sacred; or is it the process by which the DNA code of the egg or seed is made complete that is sacred; or is it the DNA code itself that is sacred; or is it the process of the development of the egg or seed into a viable living being that is scared; or is it the combination of all of the above that is sacred? As evidenced by the research put into this work, all that the Lord asked for in that regard was to love God.

The question then must be answered as to what is taken away that changes a living thing into a non-living or a dead thing. It is obvious that it is not the egg involved initially in starting the new life since that process has been completed. It is obvious that it is not the sperm that completed the DNA code so that the egg would be made active since that process was completed. It is not the process of development of the egg into a fully developed being since that process has been completed. The answer is found in a preceding section entitled "Beginning Of Living Organisms". Therein it was shown that a living cell was made up of an assembly of precise complex self-governing vital assemblies and sub-assemblies, powered through the proton/electron process, that work together under the direction of the DNA code. Since those assemblies and sub-assemblies are vital to the establishment of life, a cell would not, could not live if those assemblies and sub-assemblies fail or do not function properly together. So it is with a living being such as a human, which is the assembly of many vital assemblies and sub-assemblies or organs. A being would not be able to live if or when one or some of those vital organs fail and therefore that being would die. So, to answer the question, the only thing taken away in the process of death is the viability of the assembly that descended from the viability inherent in the original cell by design.

Conclusion? The initiator of a new life is a life viable egg cell descended from the initial creation, which new life lives until its viability is lost to produce death. So what makes a human life sacred? Could it be that the concept of the sacredness of human life is embedded in the conscience of man because of man's ability to love God, meaning willing the success of God's objective, creation? You decide.

Chapter Twelve: Alpha To Omega

Note in explanation: If as evidenced, all things seen and unseen in the Universe were created, then it stands that the energy of which all things seen and unseen were made is also a product of creation. It is the purpose of the material in this chapter to endeavor to give insight into the nature of that energy which is fundamental to understanding creation and the Creator.

The Alpha

The argument has been made that elemental particles of matter used in making the Universe (electrons, protons and neutrons) are physically and behaviorally eternally precise. The argument has also been made that i*t is the behavior characteristics of the energy of which all things seen and unseen in the Universe are made that establishes the behavior characteristics of everything in the Universe.* It is the same energy of which all physical matter is made that provides all of the forces involved in the creation of the universe.

Being divine in nature with no physical properties and therefore incapable of being examined directly, the behavioral characteristics and nature of the energy of which all things seen and unseen are made, including the four fundamental forces of nature that puzzled Einstein, are capable of being visualized and analyzed through observation of the behavior of the things physical that are made of that originating, call it universal, energy or of the behavior of things physical as they are acted upon by the four fundamental forces of nature. Whether such analyses produces new understanding or only proves existing understanding to be flawed, the results will be more meaningful than simply ignoring the fact that a source of energy is responsible for producing those fundamental forces. The important thing to remember is that the behavioral characteristics of the fundamental forces of nature and the elemental particles are built-in. **They behave as they do because that is the way they were built to behave.**

It bears repeating; *the behavioral characteristics of the originating energy (call it Universal Energy) of which all things seen and unseen are made, including the four fundamental forces of nature, is absolutely unerring and eternally precise*. This is seen from the behavioral characteristics of the force called gravitational attraction and the physical and behavioral characteristics of electrons, protons and neutrons that are made of that energy. Since the behavioral characteristics of electrons, protons, neutrons and gravitational attraction are derived from features inherent in UE (Universal Energy) those features are capable of being visualized and analyzed by observing the behavior of electrons, protons, neutrons and gravitational attraction. These are the premises and the process that are applied in this analysis.

Basic Neutron Characteristics

As established by Particle Physicists and documented in Physics text books, neutrons have a mass of 1,836 (some say 1,840) times the mass of an electron, the smallest particle of matter to exhibit physical properties (mass), which are, as reported, produced by ejection from a neutron. On that basis, since electrons are rings of energy, it is concluded that neutrons are essentially an assembly of 1,836 electron-size ringed segments of Universal Energy that are held together by the force of gravitational attraction, which as previously argued, is applied through "umbilical like" links to a source of UE. Recognizing that the Universe has been made in "building block" fashion where the building blocks are eternally precise, and that neutrons become protons after the ejection of an electron, it becomes self-evident that neutrons and thereby *all matter in the Universe are assemblies of rings of a substance that will subsequently be shown through examination of electrons to be identically precise segments of the force that produce gravitational attraction.*

It is important to recognize that while the rings of UE (Uerings) of which neutrons are made have an electron like composition, they do not have the behavioral characteristics of electrons. Given the stated explanation where neutrons are made of a precise quantity of like elemental assemblies of UE which when ejected acquire the behavioral characteristics of an electron, it must be concluded that Uerings are in fact building blocks in the process of creating matter and that neutrons play a role in that process. It bears repeating. The ringed segments of UE of which neutrons are made, while electron size, are not electrons since they have not been ejected in the process of forming a proton and therefore are not party to the proton/electron relationship known as the producer of electric energy. That is to say, until ejected from a neutron, there has not been an "umbilical like" link established between a uering and a proton, which is the mechanism responsible for establishing that special proton/electron relationship. Stated in other words, it is the "umbilical" link between a proton and an electron that occurs in the process of being ejected from a neutron that makes a uering an electron. It is also for that reason that neutrons are said to have no electrical charge or that they are electrically neutral.

The process by which a neutron "morphs" into a proton and an electron has not been conclusively established. However, it appears to occur when the strength of the gravitational attraction to which subjected is sufficient, the assembly of ringed UE segments is constricted to the point where all but one of the ringed segments of which made, are compressed into one common "center of gravity" point. Since the scenario as described defines the result, there is every reason to believe that the "morphing" of neutrons occurs in the manner described. The important point is to recognize that the neutron "morphing process" is always eternally precise producing two (the proton and the electron) of the three elemental particles of which the Universe is made, as well as the generator of the mechanism used to produce three of the four fundamental forces necessary for creation of the Universe. Not only is there no other explanation for the origin of these processes and their products, the development of these processes at such a critical point in the creation process illustrates beyond any possible counter argument that the process of creation is the work of divine wisdom. The results produced by the neutron "morphing" process fulfilled a precise purpose for which there is no other means for producing those results and creation of the Universe would not have been possible without this process. *Let it be understood, there would be no electrons or protons and no proton/electron relationship with which to assemble atoms of matter and the Universe if there had not first been neutrons with the ability to "morph" into protons and electrons that possessed that special proton/electron relationship.* There is no "which came first the chicken or the egg" in this situation and there can be no question but that the neutron "morphing" process was conceived before creation was begun. That is, before there was a first proton or electron or the force generating process needed to hold elemental particles to form atoms. ***Without the eternally precise neutron "morphing" process, there would be no creation.***

With a center of gravity established, the remnant neutron material is endowed with the behavioral characteristics that make that material a proton. However, in addition to the role described, neutrons serve another purpose. Neutrons, in their role as a neutron, when joined with protons as when forming the nucleus of atoms, provided additional mass and thereby UE strength at that "center of gravity" point which is essential in the formation of a variety of atomic formations as well as both the weak and strong nuclear forces.

Basic Proton Characteristics

Because of the way they are produced, protons possess the unique ability to function as a center of gravity. With that ability protons serve as a center of gravity for the energy involved in gravitational attraction as well as the anchor for attracting particles. It is that unique characteristic that endows protons the power to function as the foundation upon which atoms are built. In that capacity they are the anchors for electrons that are involved a myriad of atomic functions too numerous to permit addressing at this time but principally those involving electro/chemical activity. It is that unique characteristic that allows a variety in atomic configurations so that atoms possessing a variety of characteristics is possible.

That unique characteristic of the proton deserves special consideration. Before "morphing" from a neutron, the neutron had 1,836 units of electron-size elemental particles that were supplied 1,836 units of UE delivered through 1,836 "umbilical like" links of UE segments that extended beyond the neutron assembly to a center of gravity source. Therefore, the strength of the force applied in attracting those elemental particles diminished with their distance from the supporting center of gravity. When the center of gravity was established in the process of producing a proton, the "affinity" force (mass) of 1,835 of those particles plus the energy of the 1,836 "umbilical" links (that are not attracted by gravity so do not register as mass) thereby concentrated at full strength into the proton center of gravity. Therefore, even though the body of the proton embodies one less elemental uering particle (less mass) than does the neutron, the proton with 1,836 units of full strength UE at its center of gravity plus the affinity power from its 1,835 elemental particles provides the power needed to attract a neutron adding to the power of the proton center of gravity. When united with a neutron as in a nucleus, the energy of the proton/neutron assembly is 1,836 units (from what were proton umbilical links) + 1,835 units (from the mass of the proton elementary particles) + 1,836 units (from the mass of neutron elemental particles) or 5,507 units and it is that combination of power that is known as ***"the strong force" that holds the nucleus of atoms together, the fourth of the four fundamental forces of nature.***

Basic Electron Characteristics

While electrons are the smallest elementary particle exhibiting physical characteristics, their extensive involvement in the activities of "nature" has resulted in considerable scrutiny of their nature and behavioral characteristics under a variety of applications.

As was previously explained, electrons are "born" in a process where neutrons expel a sub-elementary particle used in their assembly, thereafter to be known as an electron. In the process of expelling the electron, a center of gravity is established in the remaining neutron material then called a proton that takes on completely new behavioral characteristics. ***It is the center of gravity*** that gives protons the mechanism for setting up a special unique attraction relationship with electrons that is separate and apart from the basic gravitational attraction (affinity) force mechanism. *This new unique bonding link between protons and electrons, **and protons and electrons alone**, provides the force used to hold electrons in orbit* *around a nucleus of an atom* ***(one of the four fundamental forces of nature called the weak force)*** *plus this unique proton/electron relationship establishes* ***the mechanism that forms the basis for all electrical and chemical activities.***

Study of the electron and electron behavior has revealed considerable insight into the nature and characteristics of the material of which electrons are made and thereby the force known as gravitational attraction. Briefly, electrons have been found to be made of a precise size "string" like segment of UE formed into the shape of a ring where the energy of which made travels at high speed around the ring (186,000 mps) while the "string" itself spirals at high speed and while infinitely elastic, a force exists, mirroring the force of gravity, tending to constrict these rings to the smallest possible diameter. ***The electron rings are the mechanism that produces a force known as magnetism where the individual lines of the electron rings are then called magnetic lines of flux.*** In that form, the direction of the force produced by those lines of magnetic flux is set by the direction the ring material spins as well as the direction the ring material spirals (spiral is counterclockwise when viewed from the source of the UE). Both of those directional characteristics are set by the direction electrons are being caused to move relative to the controlling proton. It is important to understand that the electron rings are not exposed except or unless the electrons are moving from atom to atom as in the flow of electrons in a conductor.

Studies of the electron and magnetism have established that the electron rings are one and the same as those called magnetic lines of flux. Since those lines are in fact a constituent part of an electron, it must be concluded that they are electron specific. *That is, one electron, one line of force with no combining, canceling or exchanging with other electrons.* For that reason these lines act to repel each other to maintain that isolation. It is that characteristic that allows the force of these lines to act in unison when directionally aligned alike but in opposition when not directionally aligned alike. It bears repeating. ***The lines of magnetic flux are in fact the electrons even though they are infinitely elastic. However, the point at which the link is established with a proton remains linked to the controlling proton in the nucleus of the atom of which a component.***

The exposed electron rings reveal that the material of which electrons are made mirrors the behavior characteristics of the lines of UE that produces gravitational attraction force precisely with but one exception. The UE involved in gravitational attraction are straight line segments of UE connecting a source of UE with an electron-size elemental particle while the segments of UE involved in making an electron and the electron-size rings of which neutrons are made, are formed into a closed ring with the "umbilical link" to the UE source attached at the point where the ends of the UE segment join to form the ring.

Basic Gravitational Attraction Characteristics

The force called gravitational attraction is also one of the four fundamental forces of nature. Gravitational attraction acts radially from the center of gravity of the Universe and of protons with the force pulling towards the center of gravity. Electrons are linked to the center of gravity of a proton, which in all cases is in the nucleus of an assembly called an atom. Electrons do no have gravitational power with which to attract even other electrons. Neutrons also do not have gravitational attraction power with which to attract anything, not even an electron. From this it is self-evident that only protons possess the means for serving as a center of gravity for gravitational attraction but all particles, including protons, are connected through a gravitational link to a center of gravity. In a body involving numerous protons, the protons act collectively as one unified center of gravity.

By analyzing the chain of gravitational attraction where the smallest particles are attracted to larger and then successively larger assemblies in succession with successively stronger and then stronger centers of gravity, it is self-evident that the source of all gravitational attraction is the center of gravity at the center of the Universe. It is from that point that the energy that powers gravitational attraction is distributed and it is that point that is the anchor for the Universe wide gravitational attraction system. The Gravitation Attraction System embodies two functions. One is to distribute a supply of gravitational force energy to each center of gravity through out the Universe. The other is to apply gravitational attraction force. Each center of gravity is supplied with one unit of gravitational attraction energy for each electron-size elemental particle attracted by that center of gravity.

Each divine in nature electron-size elemental particle that has no physical component but which are constituent components of all physical matter, are connected to their supporting center of gravity by means of one of the encased segments of UE of which all things are made (that are divine in nature with no physical component) which serves as an umbilical like link by which gravitational force is applied. ***The force, divine in nature with no physical component, directed at pulling physical particles or assemblies of particles (such as the earth for example) toward its supporting center of gravity (the Sun), is applied through those links***. One link one electron-size particle. As the Universe expanded, centers of gravity of smaller assemblies were linked in that manner to the centers of gravity of larger assemblies. For example the moon is linked to the earth, the earth to the sun and the sun to the Milky Way, etc. Stated in other words: Beginning at the center of gravity at the center of the Universe, there is one separate and independent umbilical link of UE for each electron-size particle in the Universe and between centers of gravity there is one umbilical link segment of UE for each electron-size particle supported through that center of gravity and so the line of umbilical like links continue until it reaches an elemental electron-size particle. This assures in every case that the strength of the attracting force at each center of gravity is one unit of strength for each electron-size elemental particle supported by that center of gravity.

The strength of attracting force applied by each such link diminishes with distance from its supporting center of gravity. That is, as a segment of UE is stretched to reach a greater distance, the **attracting** strength at the distant end is diminished. Those who have investigated the matter have established through experiment that the strength diminishes at a rate that is inversely proportional to the distance from the supporting center of gravity squared. That phenomenon can be demonstrated with a permanent magnet where the attraction is strong at close distance but rapidly diminishes as the distance is increased. The reaction is analogous to drinking with a straw that is uniform in porosity. The suction power diminishes as the length of exposed straw increases.

The energy delivered to the center of gravity of each proton and each intervening center of gravity, gives each of those points the power with which to establish an interconnecting "affinity" link between all atoms of matter. That link between all atoms of matter in the Universe is not only seen as causing the moon to produce tides on the oceans of the earth for example, even though the moon is satellited to the earth, but those "affinity" links establishes a web of UE links between protons throughout the Universe. The Universe wide web established by those "affinity links" forms ***a non-physical medium over which Electro-Magnetic Forces*** *(which includes light)* ***are radiated which explains how the second Fundamental Force of Nature works***.

Whether serving as a gravitational attraction link, as an electron, or as an elementary particle used to make a neutron, all such components are precisely identical segments of UE. When any segment of UE is ruptured or disconnected from it's source ***the contained energy (UE) is released as what has come to be known as heat***.

A basic characteristic of UE whether acting as gravitational attraction or as a line of magnetic flux (which is the exposed ring of electrons) or as the force in the proton/electron relationship, is that the applied force acts at the "speed of light" which has been calculated through experiment to be approximately 186,000 miles per second.

Basic Characteristics of Light

. At this point it is important to recognize that *light is not light without eyes to see.* What is seen as light by the eyes is in fact an impression made by intact segments of UE that have been released when the orbit of electrons are disrupted. The intact UE segments (either an electron or the link of UE bonding the electron to a proton in the nucleus and now called Photons) carries a sign wave corresponding to the power involved in holding the electron in orbit. As previously argued, photons are attached to and transported by the inter-particle web of "affinity" links of UE. At this juncture, the point is that an intact segment of UE is ejected from an atom and transported radially from that point at the "speed of light" in a form called photons. At some point, those segments are fragmented and the energy of which formed is released in a form called heat.

A second characteristic attributed to light, because it is revealed when analyzing light, is that light travels at 186,000 miles per second, which is customarily referred to as "the speed of light". That characteristic has been quite thoroughly investigated and proven, however; it has not been established how or why light travels at that speed. In considering this characteristic, it is of interest to note that the "speed of light" is also a characteristic involved in electrical energy, magnetic energy and gravitational attraction (especially those affinity forces that exist between distant particles of matter) that are direct applications of UE. When it is recognized that the universe wide inter-particle "affinity forces" called gravitational attraction forms a web of UE linking all particles in the Universe, and that UE acts at "the speed of light", it becomes evident that it is the web of UE that transports photons. So; ***light is not light without eyes to see and light does not travel any where without the web of UE***.

Basic Characteristics of Heat

As previously discussed, the energy commonly called heat is the energy remaining after the fragmentation of the encased segments of UE of which electrons, protons, neutrons and photons are made, or the energy in the centers of gravity or in the bonds used to link electrons, protons and neutrons. When those components are ruptured, fragmented or otherwise disassociated from their center of gravity connection, that energy is released. Said in other terms, ***heat is the UE released when those things seen and unseen that were used to make the Universe are fragmented.*** It is important to realize that our interpretation of the term heat derived because a sensation is produced by the brain when the residue energy of fragmented matter interferes with the functioning of atoms and thereby the cells of which the body is made, causes the generation and transmission of a nerve impulse interpreted as heat or burning. When atoms are fragmented, some of the encapsulated segments of energy of which the components of atoms are made are ruptured and release the energy of which made as "heat" and light photons. In some cases the released energy may be of such magnitude as to bee seen as combustion or even to produce an explosive effect. In most cases both heat and light are emitted simultaneously hence the appearance of flames. *By observation of UE released as combustion, it is apparent that UE is not affected by gravity,* which confirms that UE residue is no longer a viable particle with an umbilical link to a center of gravity. A further observation reveals that the production of what we know as heat requires in all instances, the sacrifice of the material of which the Universe is made whether in the form of oil, wood, "atomic energy", or electric current.

When physical matter and the links of the gravitational attraction system are fragmented, the energy of which constructed expands upon release and migrates towards a location where the density of released energy is less. The dispersion of released energy is accomplished through absorption by viable intact matter. In the process, the freed energy interferes with the functioning of atoms in the absorbing material causing "expansion" with the disruption of electron orbits, and the fragmentation or failure of those electrons. Given the proper circumstances, damage to cellular structures, combustion and even explosion may occur. Said in other terms, the **residue** of the energy of which matter is made and that produces gravitational attraction, is "heat" wherein "heat" is what remains of the Universal Energy after release from its encapsulated segments. Once released, as stated in the Humpty-Dumpty nursery rhyme, no one can put Humpty-Dumpty together again. That is Omega.

Summation

The nature and characteristics of the one Universal Energy of which all things seen and unseen are made as revealed by examination of the nature and characteristics of gravity, magnetic lines of flux, electro-magnetic forces, oxidation/combustion and heat are summarized as follows:

- The elemental particles of physical matter and the forces employed in their assembly are made in building block fashion where the building blocks are eternally precise identical segments of divine in nature UE with each segment individually encapsulated by a means with no physical presence, (spiraling motion?) and where the force produced acts at what is termed "the speed of light". That is the Alpha.
- The encapsulated segments of UE are employed both as the "umbilical" links in the application of gravitational attraction and as the electron-size rings (Uerings) used to make neutrons and thereby protons and electrons.
- As Uerings, the UE segments are made into the form of a ring with a separate UE segment as the "umbilical" link attached at the point the ends of the Uering segments are joined. There is one "umbilical" link for each Uering. Each Uering consists of a segment of UE formed into a ring and a second segment of UE that serves as the "umbilical" link.
- UE segments, while possessing no physical component, behave in a manner similar to sucking on a straw where the straw is precisely porous and infinitely elastic so that the force applied is directional and the strength of the applied force diminishes with the length of the straw as dictated by the porosity of the straw. In the case of "umbilical" links, it is the strength of the force remaining at the end of the straw that is involved. In the case of Uerings, since they are formed in a ring closed by the connection of an "umbilical" link, it is the strength of the force working to constrict the ring diameter that is involved.
- The encapsulated UE segments are eternally precise. They will not unite or join other encased

segments of UE and so repel each other although in "umbilical" links they work in unison when the Uerings to which attached are assembled in a common body. In the case of Uerings, their manner of behavior is demonstrated using a magnet and iron filings as commonly shown in High School Physics classes and books. In the case of the magnet demonstration, it is the lines of force of electron rings that are involved in producing the reaction on the iron filings. The point of "umbilical" link connection to the ring remains associated with a proton in the nucleus of an atom of the magnet.

- When the integrity of atoms of matter is disrupted, some or all of the segments of UE involved are released. When rings of UE (electrons) are released intact from their "umbilical" link, some are picked up by the UE affinity links emitted by the support protons and radiated as Photons of light. When the encasement of uerings or "umbilical" links are ruptured or fragmented, the contents expand to be absorbed by other intact atoms of matter to be seen as heat, which then interferes with the functioning of those atoms. Exposure to residue UE (heat), may eventually lead to failure of the exposed atoms and in turn the release of more "heat" possibly leading to a chain reaction in the form of combustion or explosion. The process as outlined encompasses activity principally involving electrons such as occurs with electric heating elements, light bulbs, oxidation and chemical activities. When the ruptured matter includes protons or neutrons, the magnitude of released UE material can be significant as in nuclear explosions.

Conclusion

Universal Energy, the E in Albert Einstein's $E = Mc^2$, the one divine in nature energy of which all things seen and unseen are made, is supplied from a source at the center of the Universe in the form of eternally precise encapsulated segments with no physical component; that spiral at a very high rate; that exerts a constricting force upon itself; which, as an extension of the material of which matter is made, acts to pull on that matter in the manner called gravitational attraction and/or electrical energy (where it pulls on electrons); and which when released from its encapsulated form, decays to return to the form from which originally made, the one divine in nature energy of which all things seen and unseen are made. That is the Omega.

After Thoughts

In God We Trust

The behavior influencing mechanism activated by the covenant offered by Jesus was not something newly created for that occasion. All of the mechanics employed by the Jesus system are real. *The influencing of behavior is made possible by biological features that were embodied in corporeal human beings at the time of creation and function whether in support of secular societal driven behavior or religious faith driven behavior.* Those features are vital to the sustenance of viable human societies, which as was previously detailed, the objective of the creator and those features were embodied in humans before there was anything called religion. Without the behavior influencing mechanism human behavior would be worse than that of wild animals since wild animal behavior is at least kept under the control of inherited instincts and personality characteristics while in human beings, inherited instincts and personality characteristics can be over ruled by the freewill that proceeds from the higher intellect they possess.

It is of interest to note that it was recognition of those characteristics with which humans were endowed by our creator, that served as the philosophy employed in the Declaration of Independence and the writing of the Constitution of the United States and it is that philosophy that is responsible for the freedoms and prosperity that have resulted. The very disturbing realization is that humans repeatedly lose the incentive to remember and embrace this truth: *the success of God's creation, and of the government of the United States, rests upon the ability of the described behavior control system to work.* It only makes sense to give credit to the originator of those endowed features that serve as the model for the philosophy embraced in the formation of the government of the United States of America with the motto: In God We Trust. It is fitting and proper for Americans to recall that it is the acknowledgement of a Creator that makes the Jesus System of Behavior Control work and it is the freedom that proceeds from the Jesus System of Behavior Control that makes The United States of America great.

Show Me God.

After all is said and done, the question remains and is frequently raised asking for an explanation of where God or a creator came from because they too must have had an origin. That is a legitimate question and any process that purports to reveal the truth about creation should produce a meaningful answer for that question. Of course the most usual answer is that the creator whether as God or otherwise, is eternal so had no origin but that answer evades the question that legitimately begs for an answer.

Part of the problem stems from the fact that the classification of the creator as God came into use as a religious term only after human beings came onto the scene when the creator then called Lord entered into a faith contract with Abraham[43] and the descendents of Abraham. Prior to Adam and Eve, and until the time of Jesus for those not a descendent of Abraham, there was no one to call the creator their Lord or God but there was a creator. Further, the identification of the creator as God even to this day occurs only with those people who believe that the creator is God. However, the fact that some people do not believe that there is something seen or unseen they are willing to call God does not change the fact that creation had a beginning or that the process of creation involved the application of unlimited wisdom. It also becomes evident that we humans have gained and continue to gain understanding of the wisdom applied in the creation process, and thereby our understanding of God, through our investigation and study of that which the Lord God created (which includes the Scripture). One of the things we have come to understand through our studies is that something that has not been revealed in some manner is impossible to understand. So it is that the full nature of God the Creator has not yet been revealed to the extent that we are able to understand fully. That does not mean that there is no Creator God. Again, that is not an answer to the question.

---

[43] Genesis 17

Our understanding at this time may not provide a fully satisfactory answer to that question but we do understood that creation involved the application of unlimited wisdom, some of which understandably, for the reason stated, we currently are incapable of understanding. The point being made is that wisdom can exist, it is, but **our** wisdom is limited to only what we understand. Wisdom does not necessarily require a **being** so that it can exist, whether spiritual or otherwise. *Wisdom can be contained within that which wisdom has created whether or not revealed*, and Wisdom might claim without fear of contradiction that I am that I am. From that realization, it should not be difficult to recognize that Wisdom that has no limits could and would conceive on its own, a way to make its existence known. At this point, the precise process may not be known, although it can be rationalized that Wisdom figured out how to produce a supply of the energy to use in the creation of the universe. There are some things that are known about the energy that Wisdom created. That energy known to produce gravitational attraction (Universal Energy as explained in All Thing, Seen and Unseen, 1stBooks © 2002) exists in spiritual form without any physical component but yet has the power to produce force on physical matter. Also, that energy has extremely powerful constricting or compressing characteristics and when extended from its source in the form of gravitational attraction, neutrons, protons and electrons, that energy exerts a force so constricted that physical properties are produced. That is, the force is pulling as though trying to pull everything to which attached, back into its source. By that action, electrons, protons and neutrons are constricted so tightly that they acquire a physical property. It is also known that when that energy is released due to the fragmentation of protons, neutrons, electrons, photons, and gravitational attraction, that energy expands to release what we sense as an expansion or explosive force releasing what we sense as light and heat. But the energy released is not really light and heat since light is not light without eyes to see and heat is not heat without senses to sense. The perception of those phenomena rests with the capabilities inherent in the eyes or senses. That is not to say that release of that energy seen as heat and light does not inflict damage to or destruction of other electrons, protons, and neutrons along with the release of their energy.

Then to answer the question where does God come from? As revealed through this analysis, God the Creator is Wisdom that exists eternally because it is. There is no energy, no form, no space, no "being" or "thing" involved. That is a plausible concept.

Show Me Satan.

For all things there was a beginning. So it is too with Satan. Satan did not exist until human beings, with intellect capability sufficient to take control of behavior away from inherited instinct and personality characteristics, were created. As a means of influencing behavior, human beings are endowed with a heart that is their conscious wherein **beliefs** are "written", which are then called upon at will in the course of making behavior decisions. Since Satan is an influence that acts upon the conscience to alter established pro-God **beliefs** or to create new anti-God **beliefs**, it is self-evident that the influence called Satan came into existence along with the heart (or conscience) faculties created in the bodies of Adam and Eve. Since Satan functions only as an influence and only within the hearts of humans, that influence would more appropriately be called Humanism that employs Relativism (that alters the meaning of beliefs) as one of the mechanisms evolved in that process.

There are a number of Scripture passages that provide insight into the nature of Satan. Jesus is recorded as having told Peter: "Get away from me Satan! You are a dangerous trap to me"[44] when all Peter did was comment in response to word that Jesus would be made to suffer, die and then he would be raised on the third day. On this occasion it was the temptation inherent in Peter's comment that Jesus classified as Satan. On another occasion when explaining the parable of the farmer scattering seed, Jesus likened the seed falling on hard ground to his message having been heard, "but then the Devil comes and steals it away and prevents them from believing[45]." On this occasion it was **belief** in Jesus that had been heard and registered in the conscience **that the Devil "stole"** before that **belief** could become

---

[44] Matthew 16: 23

[45] Luke 8: 12

established. There are numerous other occasions where mention is made that Satan and his demons occupy the heart, which has been shown to correspond to the conscience, which itself is a function of the mind.

There was an occasion as reported in Matthew 12:24-30 where, in responding to the charge that he got his power to cast out demons from Satan, Jesus responded "Anyone who isn't helping me opposes me, and anyone who isn't working with me is actually working against me." *We work for the good of God when belief in God is embedded in our conscious and we work against the good of God when belief in God is not embedded in our conscious.* When belief in God is not embedded in our conscious then God's presence is not available to influence our behavior in which case our minds are free to direct behavior according to its will without the benefit of God's presence in our conscious. The absence of the presence of a belief in God in our conscious thereby leaves our behavior to be influenced by selfish personal desires such as were inherent in the temptations Jesus rejected during 40 days in the wilderness. Therefore, could it be that *satanic beliefs, that are in fact Satan*, take over and occupy the conscious of an individual in the absence of the influence that proceeds from the love of (willing the good of) God? Could it be that Satan's influence is advanced when Relativism is used to modify the truths inherent in belief in a creator or the Commandments given by the God of Abraham, Isaac and Ishmael, or the teaching of Jesus? There is good reason for such a conclusion.

The Mystery of all Mysteries

When it comes to getting an individual or individuals to believe in a Creator or God, the first objective, as recognized by the assemblers of The Holy Bible, is to get individuals to recognize, understand and acknowledge that all things that exist had a beginning, that creation is a truth, and that a Creator produced that truth. That is why the story of creation was put at the beginning of the Bible and the opening sentence of the Nicene Creed. **But**, *acknowledgement that there was a beginning and a creator does not of itself define the beginning nor the truths and the abundance of wisdom embodied in the creation* *process*. That is the point that the Theologians and Ministers of the Judeo/Islam/Christian Religious Communities are failing to recognize when they make their stand. It raises the question as to why these people are closing their mind to that abundance of knowledge and understanding. Do they despise the wisdom contained within the work of the Creator? Is there a purpose served by collaborating to keep that knowledge hidden?

To learn is to gain understanding, which requires the assembly and analysis of bits of information so as to produce knowledge. In the process of analysis, information bits can be assembled and analyzed in a variety of ways and circumstances, depending upon the experiences, knowledge and mindset of the analyzer, with each analysis capable of producing a variety of different understandings. This phenomenon is generally understood but it plays an extremely important role in the context of this explanation.

A realization and acknowledgement that there was a beginning and a creator cannot be forced into the mind of people especially when a mind-set has been established against such belief or when the mind is filled with countering beliefs. Some people such as those seeking spiritual solace are often receptive to being convinced to trust in the belief that there is a God who is the father, the creator of all things. With such people, the "seed" that sometimes instills the search for spiritual solace is usually planted when as children they are exposed to the religious beliefs of parents. Some people are introduced to such belief through friends or through observation of the behavior of others. Some people come to that awareness through their own observations. The process of attracting people to give consideration to belief in a creator is uncertain at best but is complicated by the incessant anti-creator conspiracy campaign run by the Academe Community in collusion with the Scientific Community.

The process of convincing people to believe is further complicated by a generally accepted conclusion in Christian communities that people cannot come to an awareness that there is a Creator who is God except or unless brought to that awareness by God. The basis for that belief comes from the statements of Jesus as recorded by John 6: 44—45: "For people can't come to me unless the Father who sent me draws them to me—. As is written in the Scriptures, 'they will all be taught by God'. Everyone who hears and learns from the Father comes to me." But who is to say how that is accomplished. Since we were created with an intellect in God's image, do we not have the intellect to come to an awareness of God through study of the work of God (wherein all wisdom is embodied and from which all knowledge is derived)? Would that not constitute being brought to that awareness by God? Then too, as discussed in Chapter 3, the phenomena of Relativism is certainly a factor in closing the eyes and hearts of people so that they are prevented by their own actions from understanding even though some may be professionally engaged in the study of "the work of the creator."

There are those who say that the Holy Spirit brings people to Jesus but that is brought into question by the comment of Jesus as reported in John 14:15-17. Jesus said: "If you love me obey my commands. And I will ask the Father, and he will give you another Counselor, who will never leave you. He is the Holy Spirit, who leads into all truth." Is that not a contradiction since the Holy Spirit would not come unless a person already believes and then loves Jesus?

Apostle Paul added further confusion when he made the point a little differently in his Letter to the Romans 1:19—20 where it is said: "For the truth about God is known to them instinctively. God has put this knowledge in their hearts. From the time the world was created, people have seen the earth and sky and all God made. They can clearly see his invisible qualities—his eternal power and divine nature. So they have no excuse whatsoever for not knowing God." Paul continued that message in Romans 1:25 where he said: "Instead of believing what they knew was the truth about God, they deliberately chose to believe lies. So they worshiped the things God made but not the Creator himself,—." *These statements by Paul indicated he believed that it is the people themselves who deliberately chose not to believe what they know is the truth.* But there is a severe problem with the conclusions of Paul. First: there is no evidence that knowledge of the truth about God is instilled into people as an instinct. Second: there is a lot more to what God made than the earth and sky and those things that are clearly visible.

On another occasion as recorded in Matthew 13:12—15, while explaining why he used stories as a means of teaching, Jesus explained: "You have been permitted to understand the secrets of the Kingdom of Heaven, but others have not. To those who are open to my teaching, more understanding will be given, and they will have an abundance of knowledge. But to those who are not listening, even what they have will be taken away from them. That is why I tell these stories, because people see what I do, but they don't really see. They hear what I say, but they don't really hear, and they don't understand. From these references the position of Jesus seems clear. Some people fail to learn, understand and gain knowledge because those people are not open to (interested in) his teaching and are not listening. There is no indication that Jesus or Paul thought it was God that is preventing those people from understanding.

This mystery of all mysteries is an ages old problem that predates the message carried to the people of Israel by Isaiah to which Jesus referenced.[46] The Scripture is replete with examples of how the people of Israel closed their hearts and minds to the Commandments of God. King Solomon addressed it by advising: "Cry out for insight and understanding. Search for them as you would for lost money or hidden treasure. Then you will understand what it means to fear the Lord, and you will gain knowledge of God. For the Lord grants wisdom."[47] In Proverbs 9:10 Solomon advised, "Fear of the Lord is the beginning of wisdom. Knowledge of the Holy One results in understanding." The words of Solomon are continued in Ecclesiastes where in Chapter 2:26 he says, "God gives wisdom, knowledge, and joy to those who please him." It was Solomon's concluding advice to: "Fear God and obey his commands, for this is the duty of every person."[48] Look again at what Solomon said! When you search for insight and understanding you will gain knowledge of God who is the source of wisdom.

Therein lies the mystery of all mysteries. It is known from previous argument that the Creator God does not force people to

---

[46] Matthew 13: 12-15
[47] Proverbs 2:3 - 6
[48] Proverbs 12:13

learn. Instead God endowed humans with an intellect in his image with the ability to reason under the control of freewill so that humans have freedom to control what is learned and what beliefs are established in their hearts as a result of that knowledge. That is why it is a mystery why some of the most educated humans find it impossible to ascribe to belief in creation and a Creator, even though their work in some cases is the study of the creation process. What is more, why do they conspire to obfuscate and hide indisputably true facts that they themselves have revealed as being critical in the functioning of the belief based human behavior control system and the education of our children? This surely is a mystery in the order of the "666" mystery in Revelations. There is but one logical explanation. Such people do not want truth to be the influencing factor for human behavior since the truth exposes their relativistic behavior and liberal agenda as the work of Satan.

Breaking the Conspiracy.

The future of mankind is in the person of our children, God's glory. They are vital in continuing the process of revealing the wisdom embodied in creation. In that regard, the education of our children is the single most important responsibility of parents. Each child develops from a single cell into a human being. When life starts at conception children have no brain. A child's brain develops cell by cell so that at birth, except for inherited instincts and personality characteristics, a child is born with a brain made of a lot of empty neuron cells. Learning, which is the absorption of knowledge, begins as soon as a child's brain is formed. A parent has perhaps twelve years at best to make sure that their child has absorbed the knowledge and beliefs needed for their behavior control system to work. Of that formative time, the first four years are the most critical since it is at the early stages of learning that knowledge necessary for healthy social activity is learned. It is during that time that the foundation of knowledge to be used in making decisions concerning right and wrong will be stored in the conscience of children. During that time children will, or should, begin to develop faith relationships that involve trusting in a belief, first with their mother then with others in the family unit but hopefully also with God the creator, the protector, the comforter. It is a parent's responsibility to guide and nurture this learning process and most do an outstanding job.

But, reverting to a common expression, "up jumps the devil". No. Not as temptation to make children engage in bad behavior but rather the devil is waiting at the front door of the schoolhouse. Is it that the devil stands guard to protect the children from the admission of something that might be hazardous or dangerous? No. The devil stands guard to enforce a prohibition, fabricated to support the relativistic philosophy against bringing into school the concept that a Creator is responsible for our existence which belief "opens the door" to all knowledge. A prohibition fabricated allegedly because that truth concerning our origin may offend Josef or Mohamed even though Josef and Mohamed or their parents, profess to hold the same belief as regards the role of a creator. The only child then possibly to be offended is one whose parents do not believe in a creator and so did not teach their children. How will they ever learn if they are not taught? But it doesn't just stop there. In class, the teachers are forced to use textbooks that teach that there is no Creator, no creation and therefore no God at all and the teachers are prohibited from allowing any challenge to those mandated position. For parents who taught their children that there is a Creator God then the school administers who teach there is no Creator, to be praised and worshiped as God, are guilty of teaching religious beliefs which is precisely what these teachers say they are not allowed to do. If teaching creation by a creator is teaching religion then teaching that creation is not the work of a creator is likewise teaching religion. Further, contradicting the parents on these matters is in fact inferring that these parents are stupid or worse. But it doesn't just stop there either. When the children get to Physics, Biology or Chemistry class their teachers and textbooks fill them with "scientific theories", some of which are not only not provable but are in fact deliberate fabrications, just to keep from having to acknowledge an origin or an originator for the existence of the universe of which we are a part. Our children are being forcefully taught false information in school under threat of failing grades and even expulsion. Worse yet, erroneous beliefs that are critical in the process used in making behavior decisions are being established in our children's conscience "data bank" by this practice. Since these are beliefs that are a vital element in the freewill decision making process our children are made to start life's journey with a behavior controlling conscience filled with anti-truth beliefs making it necessary at some point in life for them to learn or relearn beliefs based upon the truth, in many cases the hard way and in some cases after it is too late.

There are two extremely important truths that are hidden within the practice previously outlined. One: The prohibition against teaching established truths about the origin of our universe is in fact a prohibition against learning. Two: If teaching that our universe was created (as herein described) constitutes religious teaching, then it follows that the teaching of an alternative origin likewise constitutes religious teaching. There cannot be one without the other.

When an analysis of the facts pertaining to the origin of the universe leads to the conclusion that creation is the product of an "out of this world" source of unlimited wisdom, energy and intelligence, that conclusion alone should not cause that belief to be categorized as a religious belief. Such a belief becomes religious only when that source of unlimited wisdom, energy and intelligence responsible for creation is acknowledged, praised and worshiped by individuals to be their God. By the same measure, if following analysis of those same facts a person does not find that source of unlimited wisdom, energy and intelligence worthy of praised and worshiped as their God, such conclusion should not cause that belief to be categorized as not a religious belief. As long as those conclusions are based upon the same facts, beliefs developed from such conclusions do not change any of the facts pertaining to how the universe came into being nor are those facts changed because some people conclude that the creator is worthy of praise and worship as their God. Arguing over the facts pertaining to creation is another matter covered in Chapter 9.

It cannot be denied. A conspiracy exists and has existed since the time of Aristotle and Plato. In this conspiracy, the leaders of the academe and scientific communities are aligned against the Theological community to suppress the truth concerning any and all evidence of an intelligence that initiated creation and managed the progressive development of the universe and there are ulterior motives behind that conspiracy. First, the inter-action between the academe and scientific communities should be recognized since they are not really two separate communities. Most of the Research and Teaching Scientists are in fact affiliated with universities to whom they are financially invested both because universities are a source of grant money and provide laboratory facilities but also because of the salaries received as invested professors The university environment also provides a forum and a captive audience for their ulterior theories. Second, if the concept of an intelligent creator was acknowledged, the need for so much research and the prestige of Research Scientists and University Professors along with the supply of grant money would diminish. Third, it is these same academia that write the textbooks and teach the teachers that teach our children and they maintain their solidarity and status through involvement with one of the most powerful socialistic "labor" unions in the world as well as a myriad of professional organizations all intent on one thing, their own selfish personal interests. Why should our children be denied access to the knowledge such as that documented through this effort?

There is a fourth element in the previously described conspiracy. The Religious Community is complicit in the Scientific/Academe conspiracy by making their countering argument using beliefs based upon the very simplified Genesis accounting of creation that does not reveal the wisdom and knowledge embedded in the creation process and so provides no appreciative convincing argument to counter the Scientific/Academe argument. Belief through "blind" faith works fine for those who are "inclined" to believe through "blind" faith but that leaves a lot of unconvinced unbelieving people capable of being convinced to believe if the processes of creation were properly argued. As documented herein, creation cannot be understood if consideration does not begin from the beginning. The Genesis story does not provide sufficient definition for the pre Adam and Eve period thereby, the processes and knowledge involved in creation are not revealed by that accounting and the Religious Community is left as asserted, without a convincing argument. In fact the Religious Community takes the position that since they already believe in God the Creator they have nothing more to prove or learn through studying and arguing the creation process. In this instance, the Religious

Community is derelict by failing to embrace the wisdom and knowledge embodied by the work of the creator that has been exposed but erroneously argued by the Scientific Community. Even if non-believers are unable to be convinced to believe through such argument, the biased liberal agenda driven and divisive argument of the Scientific/Academe Community could be more effectively countered if forced, by such counter argument, to account in their explanation from the moment in time when the first neutron, proton and electron was made. It sure wouldn't do any harm to try even if all did was to help stem the tide of Relativism.

The Last Word

In Ecclesiastes, King Solomon repeatedly used the expression, "everything is meaningless, like chasing the wind." Also, a popular song of the Peacenik's 60's era incessantly maintained that, "The answer is blowing in the wind". In both instances, these expressions were used with reference to frustrating contradictions illustrating lack of rhyme or reason for the realities the people of the time perceived. Not much changed in the intervening or in the subsequent period.

As stated in the Introduction, the motivation for this book grew out of the realization that the truth concerning creation is being deliberately hidden and distorted through modification of the facts to accommodate preconceived biased ideological concepts or through the application of Relativism. Work on this book proceeded on the premise that if existing knowledge concerning the origin of creation was aligned in the logical order dictated by self-evident facts, freed of myth, the biases of the scientific and religious communities, traditions, and conflicting interpretations, the knowledge revealed would open the door to a fuller understanding of the work and the word of the Creator. With that fuller understanding of the wisdom embedded in creation, it would no longer be necessary to chase the wind looking for meaning or answers.

The knowledge and wisdom revealed in the course of examining the creation process in the manner described has not been disappointing. In some cases the knowledge revealed provided explanation and definition for situations already known while a number of profound revelations were exposed. Consider the following:

1. *A profession of belief, as in Jesus or God, is next to meaningless*, like blowing in the wind, *if that profession of belief does not encompass belief in and trust of the tenants of a faith contract,* such as the Commandments or the related teaching of Jesus.
2. *The wisdom that directed creation is embedded within the work of the creator. All knowledge is derived from the study of the Creator's work,* which includes the Scripture, (The Word of God). The Creator is the source of all wisdom.
3. *Freewill is a function of the provisions that are designed into the physical composition of the human anatomy.* Namely, an intellect made in the Creator's image. Freewill is not merely a religious matter.
4. *The behavioral characteristics of all things*, except behavior under the direction of freewill, *are eternally built-in.* The behavioral characteristics of all things are built out of the behavioral characteristics of the energy of which electrons, protons and neutrons are made.
5. The energy of which all things are made has no physical component yet produces a force with no physical component (seen as gravitational attraction) that emanates from the center of the Universe and by pulling upon the physical components of the Universe, holds them in place.
6. Each proton develops an affinity for an electron when neutrons "decay" to produce a proton and an electron. The proton/electron relation, known as electrical energy, is the source of energy that powers all things in the universe not powered by gravitational attraction directly.

7. Atoms of matter (assemblies of electrons, protons and neutrons) that themselves are not a living thing, when properly arranged and powered by proton/electron relationships, such as in a biologic cell, have the ability of making logical decisions to direct or influence themselves or other such non-living assemblies, to store data, and even of producing physical force. *A biological cell with all of the non-living components assembled in the manner required to be a living cell, is a living cell when so properly assembled and nurtured.* Each type cell, like each of the elemental particles, is made to possess specific behavioral characteristics.
8. *It is the biological cell that holds the secret of life. It is a functioning living cell or a cell whose functionality is made complete through "fertilization" that is the "mechanism" by which life is perpetuated from generation to generation.* All living things are descended from an "Eve" cell.
9. Relativism, the instrument of Satan, is a principal player in the denial of creation by those who practice Relativism. Those who practice Relativism do not want to face the truth that is found in the creation process but would prefer to chase the wind. Acknowledgement of the truths embodied in the creation process would render Relativism and Satan impotent.
10. *The mental capacity of humans, made in the image of the Creator, is provided with a conscience wherein embedded beliefs influence freewill decisions and thereby behavior.* Such is the mechanism that the Creator devised to influence behavior of human beings where control has been released to freewill. It is in that mechanism that the Creator has placed his faith when it comes to the success of the freewill experiment as well as the salvation of civilization. Therein also, lays the

rationale behind the Commandment to Love the Lord your God.

11. The mystery of mysteries remains from the beginning of human history until the present time. The mystery is why (especially since the advent of Jesus) are the eyes, ears and hearts of so many human beings of acclaimed wisdom closed to acknowledging the wisdom embedded in the creation process when the wisdom being denied is the product of acclaimed "scientific" investigations? Why are so many people with acclaimed intellect, hiding from the truths revealed as acclaimed scientific findings, through the practice of Relativism?

About the Author

A self-taught registered Professional Engineer (retired) who studied basic Electrical Engineering through The International Correspondence School. Freed of the thought regimen inherent in resident type institutions for higher learning, the author relied upon independent study and an inbred skeptic nature. The questioning of underlying theories and principals became the means for gaining understanding of complex matters which activity peaked in conjunction with the mass of research material published in response to the 1957/58 International Geophysical Year. Years of experience working in such places as Islamic Saudi Arabia, Buddhist Thailand and the mixed heritage in the Philippine Islands have afforded opportunity for the enriching and broadening of knowledge on a variety of subjects.

www.ingramcontent.com/pod-product-compliance
Ingram Content Group UK Ltd.
Pitfield, Milton Keynes, MK11 3LW, UK
UKHW041943190726
13854UKWH00004B/1764